5th Grade Math

5th Grade Math

Ga Standards

Lesson and Unit Plans included

Randall Aldridge

Vicki Vernadoe

Randall Aldridge
5th Grade Math

Published by Spines
ISBN: 979-8-89691-726-7

Spines

Contents

Unit 3:

Unit 4:

Unit 5:

Unit 6:

Acknowledgments

This book was written by Vicki Vernadoe. To the ones who dare to dream, I stand beside you. To the ones who struggle, I've been in your shoes.

Acknowledgments

Welcome to the 5th Grade Math: Aligned with the Georgia Standards of Excellence. This book aims to provide educators, students, and parents with a comprehensive understanding of the mathematical concepts and skills essential for fifth-grade learners in Georgia.

Purpose and Goals

The purpose of this book is to ensure students gain a strong foundation in math, enabling them to think critically, solve problems, and succeed in future academic pursuits. The goals of the Georgia Standards of Excellence for 5th Grade Mathematics include:

- Developing a Deep Understanding of Numbers: Students will explore whole numbers, fractions, and decimals, understanding their relationships and operations.
- Enhancing Problem-Solving Skills: Learners will apply mathematical reasoning to solve real-world problems, emphasizing the importance of perseverance and strategic thinking.
- Building Geometric and Measurement Skills: Students will engage with concepts of volume, area, and coordinate planes, fostering an appreciation for geometry in everyday situations.
- Fostering Data Analysis and Probability: Learners will interpret data and understand basic probability, enabling them to make informed decisions based on data analysis.

These lessons align with the Georgia Standard ELAGSE6RL1 (Georgia Department of Education, 2024).

Organization

This guide is organized into units that align with the Georgia Standards of Excellence, providing clear objectives and outcomes for each mathematical domain. The units include:

Unit 1: Investigating Volume of Solid Figures

Unit 2: Building an Understanding of Place Value Using Measurement & Data Reasoning

Unit 3: Building Conceptual Understanding of Multiplication & Division of Whole Numbers

Unit 4: Building Fraction Understanding

Unit 5: Making Sense of Fraction Multiplication and Division

Unit 6: Extending Place Value & Working with Decimals to Solve Problems

Unit 7: Exploring Geometry & the Coordinate Plane

Instructional Strategies

To support differentiated instruction to address learning styles and needs, this book includes a variety of instructional strategies:

- **Hands-On Activities:** Exploration through manipulatives and real-world applications.
- **Collaborative Learning:** Promotion of group work and paired discussions to enhance understanding and communication skills.
- **Technology Integration:** Utilization of digital tools and resources to engage students and enhance learning within the classroom or at home.
- **Assessment and Feedback:** Implementation of formative and summative assessments to guide instruction and provide meaningful feedback to stakeholders.

Our Difference: We have included time-tested strategies of review and instruction along with new perspectives to help strengthen instruction and learning. Our Unit Plans, Lesson Plans, along with Presentations give the teacher a launching pad for instruction. For the student, the Lessons and Presentations provide for learning in the classroom and at home. Parents can see what is covered and help their student as needed.

Pretest

5th Grade Pretest - 1st Assignment

1. Evaluate the following expression: Given: a = 2, b = 5, and c = 4. $5b-6c+(3a+2b)^2$. 5.NR.5.1
(Dok 3)

A. 34 B. 22 C. 289 D. 257

2. Solve for *x*. $8x = 144$. 5.NR.2.1
(Dok 2)

A. 14 B. 18 C. 13 D. 24

3. Find the volume of a cube with 3 cm edge lengths.

A. 9 cm^3 B. 27 cm^3 C. 12 cm^3 D. 15 cm^3 5.GSR.8.3 (Dok 1)

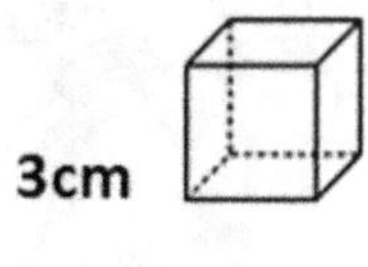

4. Find the volume of a rectangular prism with a length of 9 inches, width of 5 inches, and height of 4 inches.

A. 202 in^3 B. 180 in^3 C. 52 in^3 D. 48 in^3 5.GSR.8.4 (Dok 2)

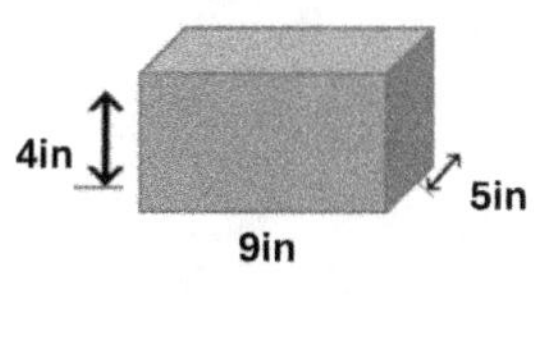

5. Evaluate (3×5) - (7+2). 5.NR.5.1 (Dok 2)

A. 5 B. 6 C. 7 D. 9

6. Find the volume of a rectangular prism that has a base of 25 square ft and a height of 9 ft.
5.GSR.8.4 (Dok 2)

A. 45 ft^3 B. 55 ft^3 C. 65 ft^3 D. 75 ft^3

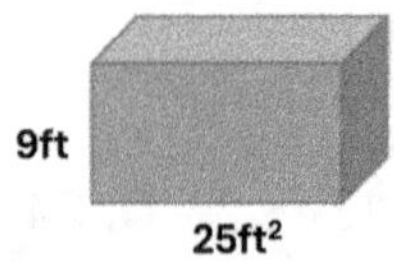

7. Drake and Casey had bought a pizza to share. Drake had eaten 1/3 of it before Casey got a chance to eat 1/4 of it. How much pizza do they have to eat for the next day? 5.NR.3.3 (Dok 2)

8. Jason was saving money for a new bike for his birthday. If the bike costs $150, and he has 25% of it, how much money does he have left to save? 5.NR.5.1 (Dok 2)

A. 120 B. 112.5 C. 125.5 D. 198

9. Sancho was explaining to his sister that the following shapes are similar. A circle, a sphere, and a cone. How are they similar, and how are they different? 5.GSR.8.1 (Dok 1)

A. Each have diameter
B. Each have height
C. Each have a radius
D. Cannot be determined.

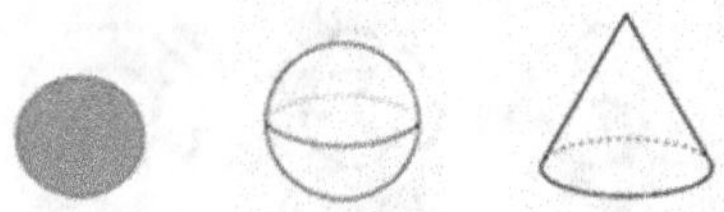

10. Kyle and Elise were cleaning their rooms before their grandparents arrived. They decided that it would be a good idea to display the model cars and horse figures on the shelves in their rooms. One horse is the size of three model cars. If a toy horse is a foot long, and a model car is 4 inches long, about how many of each would fit on two three-foot shelves? 5.NR.3.1 (Dok 2)

11. Given the number: 259,648. Which number is located in the ten thousands place? 5.NR.4.1 (Dok 1)

A. 4 B. 6 C. 8 D. 5

12. Multiply the following decimals: 9753.2 and 85.34.
5.NR.4.3 (Dok 1)

13. Jordan sees an ad for a new online game on the internet. The ad says that there will be prizes given for the 1st through 3rd place winners. The first prize is $20; the second prize is half of that, and the third prize is a fourth of the first-place prize. If Jordan considers for a moment that there are about a thousand contestants and twice as many who clicked to see the ad, should he try to win the 1st place prize or skip the ad?
5.MDR.7.2 (Dok 1)

14. Mountains in Colorado are 15×10^3 feet above sea level. Mountains in Georgia are around 20×10^2. How much higher are the mountains in Colorado compared to the mountains in Georgia? Express your answer in powers of 10. 5.NR.1.2 (Dok 2)

15. Subtract: 5.NR.3.3 (Dok 2)

$$\frac{3}{4} - \frac{1}{5}$$

A. 1/3 B. 11/20 C. 1/8 D. 2/35.

16. Karla's class is going to Lookout Mountain to see the Incline Railway. If there are thirty-three students, two teachers, and three chaperons, how many activity vans are needed if one van holds twelve people and a driver?

5.NR.2.2 (Dok 2)

A. 4 B. 6 C. 9 D. 12

17. Angel's parents bought her a used car for college. It had 76,500 miles on it when they purchased it. After graduating college, she had 265,700 miles listed on the odometer. How many miles did Angel drive her car once she got it?

5.NR.4.4 (Dok 1)

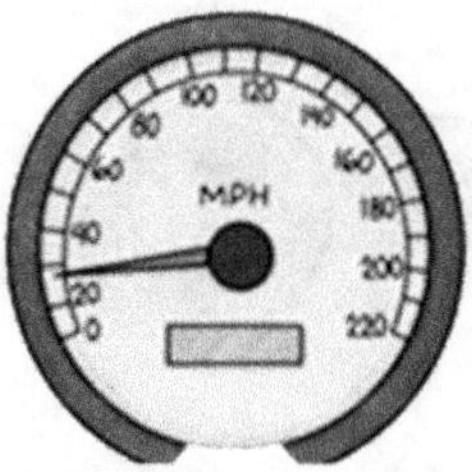

18. Drake's dad has a yard of free-roaming chickens. If there are three times as many hens as roosters, how many chickens would be in his yard if Drake had counted 8 roosters? What is the ratio of roosters to hens and the ratio of hens to chickens, minus the cat?

5.NR.3.1 (Dok 2)

19. Josh and Jacob are at a fishing tournament. The smallest fish can be 2 and a half pounds. Josh could easily qualify since his sunfish weighs 5 and two-thirds pounds. What is the difference in the weights? 5.NR.3.3 (Dok 3)

A. 3 2/6 B. 2 1/6 C. 3 1/6 D. 4 1/5

20. A recreation department received a shipment of 50 footballs, 30 soccer balls, and 60 basketballs. Which expression shows the amount of sports equipment received. 5.NR.5.1 (Dok 2)

A. 5f+3s+6b B. 10(5f+3s+6b)

C. 5(2f+3s+2b) D. -5f-3s-6b

21. Which two expressions are equivalent to 4x + 8y? 5.NR.5.1 (Dok 2)

A. 2(x+3y) B. 4(x+2y) C. 5x-x+4(2y)

22. What is the rule for the expression? 24, 30, 36, 42, 48 5.PAR.6.1 (Dok 1)

23. How many inches are in a mile? 5.MDR.7 (Dok 1)

A. 5,280 B. 60,155 C. 63,360 D. 45,205

24. Divide 3,875 by 25. 5.NR.2.2
(Dok 2)

A. 155 B. 165 C. 175 D. 185

25. Which of the following are improper fractions? 5.NR.3.2 (Dok 1)

A. 5/4 B. 1/2 C. 4/5 D. 1/9

26. Change 27/2 to a mixed number. 5.NR.3.1 (Dok 2)

A. 13 1/2 B. 12 1/13 C. 1/34 D. 5 3/4

5.NR.3.6 (Dok 1)

27. Divide 2/3 by 1/4. $\frac{2}{3} \div \frac{1}{4}$

28. Which engine has more horsepower? 5.NR.4.2 (Dok 1)

289 or 197

A. 197 B. 289 C. 92 D. Cannot be determined

29. Shelby is confused. Her grandmother sent her to the store to get 42 eggs for a meal. How many dozen eggs would she need to buy for her grandmother? 5.NR.2.2 (Dok 2)

A. 2 B. 4 C. 6 D. 8

30. Justin's parents have convinced their son that he needs a job for the summer. His mom said that she would give him $10 for the chores that he completes. Not to be outdone, his dad said that he would give him double of whatever amount of chores he completes, starting with a nickel. After a month of completing 20 chores, which offer would have benefited Justin the most? 5.PAR.6.1 (Dok 2)

31. Given the following fractions: 1/2, 2/3, 6/7, 8/9, order them from least to greatest. 5.NR.3.2 (Dok 2)

32. Given the following decimals: 0.712, 0.7246, 0.707, 0.7222, order them from least to greatest. 5.NR.4.2 (Dok 2)

33. Graph the following points on a coordinate plane.

A. (1,2) B. (2,3) C. (4,1) D. (5,4) 5.PAR.6.2 (Dok 1)

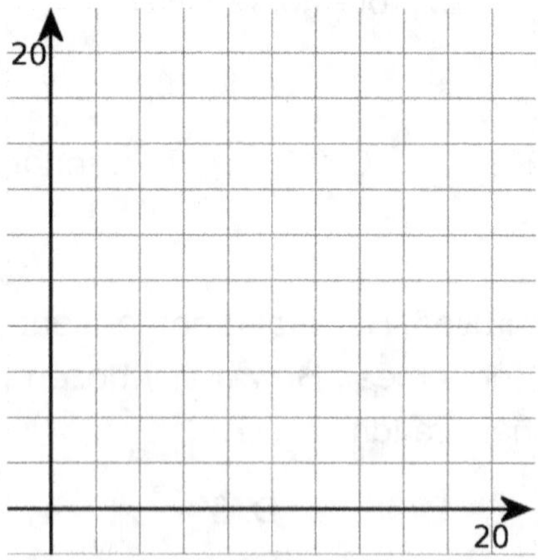

34. Which points are in Quadrant I? 5.PAR.6.2 (Dok 1)

A. (1,1) B. (-2,3) C. (-4,-4) D. (5,-1)

35. Which points are on the x-axis? 5.PAR.6.2 (Dok 1)

A. (0,1) B. (3,0) C. (0,-4) D. (4,2)

36. Find the volume of the composite figure. There is a cube that has edges of 5cm and a rectangular prism of length 8cm, width 5cm, and height of 6 cm. 5.MDR.7.1 (Dok 2)

37. Solve the equation: 3/4x = 18. 5.NR.3.4
(Dok 2)

A. 23 B. 24 C. 25 D. 32

38. Combine like terms: 3(x+2)-(x-1). 5.NR.5.1 (Dok 2)

A. x+2 B. 4x-2y C. $-x^2+4x+6$ D. 2x+7

39. Which number is greater? 365.25 or 365.18 5.NR.4.2 (Dok 1)

A. > B. < C. = D. Cannot be determined.

40. Jason has a savings account. He added $100 as an initial deposit. He wants to add ten percent of each paycheck for the next four weeks in order to buy a game. Jason earned $120 in week 1, $155 in week 2, $ 160 in week 3, and $175 in week 4.

How much money can he spend on his game if he wants to spend 1/4 of his savings on the game? 5.PAR.6.1 (Dok 2)

41. Combine like terms: 5(x-4)+3(x+7). 5.NR.1 (Dok 2)

A. 8x+2 B. 8x-1 C. 8x+1 D. 2x+1

42. Given the following decimals: 5.2, 8.1, 13.5, 0.9, 0.01, order from least to greatest. 5.NR.4.2 (Dok 1)

43. Given the following mixed fractions: 1 1/5, 2 3/4, 2 1/8, 3 4/5 , order from greatest to least. 5.NR.3.2 (Dok 2)

44. Divide: 2535 by 15. 5.NR.2.2 (Dok 1)

45. Multiply: 1872 by 23 5.NR.2.1 (Dok 1)

46. Round the decimal to the hundredths place. 883.247

5.NR.4.3 (Dok 1)

47. How many times is the digit 8 in the number 1,788 different in the tens place and the ones place? 5.NR.1.1 (Dok 1)

48. How many hours are in 3 days? 5.MDR.7.1 (Dok 2)

A. 48 B. 72 C. 84 D. 142

49. Jessie and Conner are painting toy boxes for their little nephew. If Conner can paint in half the time as Jessie, how long will it take him to paint his box if she painted hers in 3 hours? 5.MDR.7.1 (Dok 2)

A. 6.3 B. 6.1 C. 1.5 D. 1.23

50. Tristan and Luis are raising money for a charity donation. Tristan cooks twice as many cakes as Luis. If Luis bakes 19 cakes and sold three. How many cakes did Tristan bake? 5.MDR.7.1 (Dok 2)

Unit 1:

Investigating Volume of Solid Figures

Unit 1 - Vocabulary

These are the vocabulary words to be discussed in this unit. Try to have fun with them and discuss them with your learning partners and stakeholders, i.e., parents, students, coaches.

Area of a base: The surface area of a flat face that forms the bottom of a three-dimensional shape.

Cubic units in relation to lengths: Standard unit of measure to measure the volume of an object.

Edge Length: The length of two adjoining faces of a three-dimensional shape.

Equation: A mathematical statement showing equality of two expressions.

Expression: A mathematical statement containing terms and operations.

Gap: A space in between two objects.

Height: How tall an object is.

Length: How long an object is.

Liquid Volume: The amount of space occupied in three-dimensional objects.

Overlap: When part of one object covers another object.

Right rectangular prism: A three-dimensional shape having faces, edges, and right angles.

Solid figure: A three-dimensional object or shape.

Unit cube: The customary unit of measuring the volume of an object.

Volume: The amount of space occupied by a three-dimensional object.

Create a story using the vocabulary. Try to make the story about a page in length. Share it with friends.

Unit 1.1 5.GSR.8.3 Plan into Practice (DOK 1)

Area and volume are concepts that we use often. Imagine you have been invited to go look at houses by a family member who is considering moving to the town you live in. The first home is an A-frame home with about 1,200 square feet of living space. The yard is inviting in the front. It has a little room to park two cars, but the backyard is an acre of woods. The second home is a two-story older home with 4,800 square feet of living space and 10 acres of yard and/or pasture. It has a small barn with a creek on the edge of the yard.

The first home and acreage lists for $120,000.

The second home and acreage lists for $500,000.

Which would you think is a great buy and why?

Yes, area helps us to know how large a 2-dimensional surface is. While volume is more concerned about the space occupied by a 3-dimensional object.

Let's consider finding the volume of rectangular prisms.

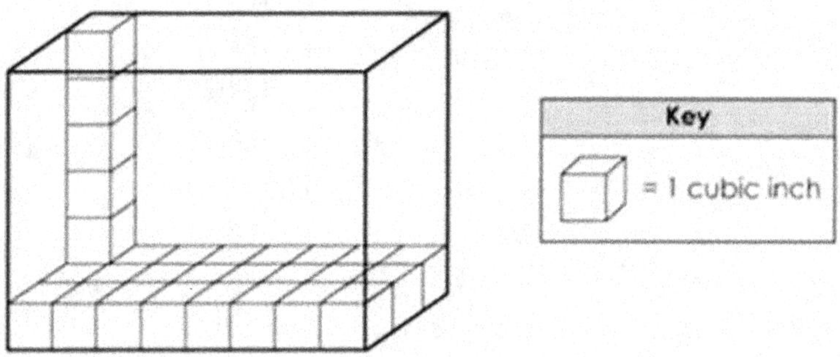

Example 1: Find the volume of a rectangular container with dimensions 3 units by 4 units by 5 units. We will use unit cubes to fill in the space within the rectangular prism. First, we will place unit cubes in the bottom of the container, and then we can count the number of units. This gives 15 unit cubes in the bottom of the container. Next, continue to fill the rectangular container with unit blocks and continue counting them. We should count 60 unit blocks for this container.

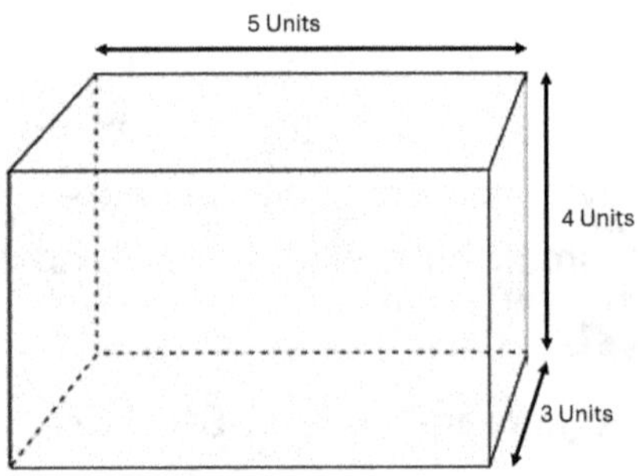

Example 2: Another way to find the volume of a rectangular prism that has a length of 5 units, a width of 2 units, and a height of 4 units is to count the blocks as you create the rectangular prism being made. We should count 40 blocks for this rectangular prism.

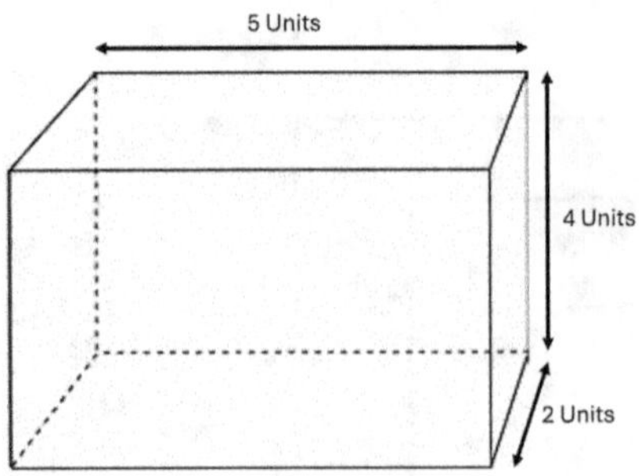

Example 3: Your turn: Determine the volume of a rectangular prism that measures 4 units by 3 units by 5 units. You should have counted 60 unit blocks for this rectangular prism.

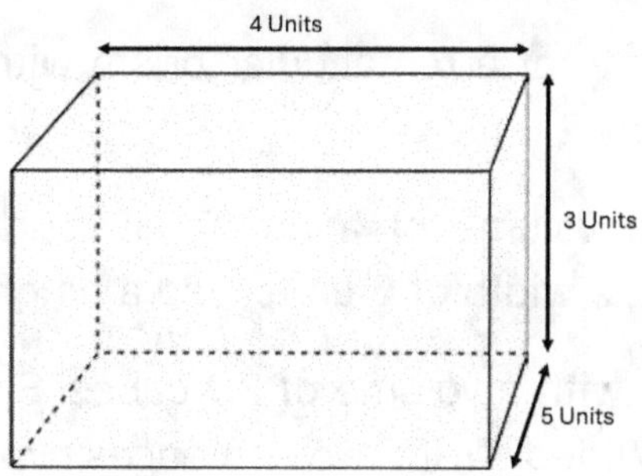

Example 4: Consider the pattern of a graph created by the volume of a rectangular prism. For example, we find the area of a base, and we calculate a base of 6. We add a layer, and the number of cubes is 12. We add another one. How many cubes now? 18. What kind of relationship is created by plotting the points? This should create a linear relationship.

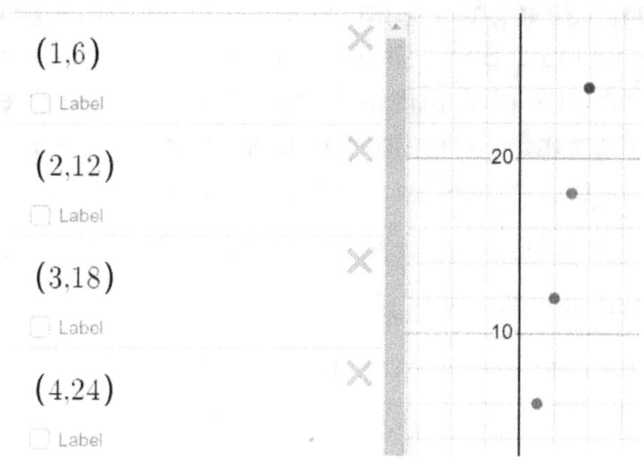

Exercises:

1. Find the volume of a rectangular prism with dimensions 4 units by 3 units by 2 units.

2. Find the volume of a rectangular prism with dimensions 5 units by 6 units by 2 units.

3. Calculate the volume of a rectangular prism that has a length of 5 units, a width of 6 units, and a height of 3 units.

4. Build a prism with a volume of 30 cubes. Sketch the figure in the space provided. Write two sentences explaining why you think it has a volume of 30 cubes.

5. Build a prism with a volume of 45 cubes. Sketch the figure in the space provided. Write two sentences explaining why you think it has a volume of 45 cubes.

6. Build another prism with different dimensions with a volume of 45 cubes. Explain how you know that the volume is the same as the first prism.

7. Your Math team has been chosen to represent your grade at the Math Quiz Bowl. Write a 4 sentence note explaining what the volume of a prism means. Give a specific example different than those mentioned above. The best explanations get to move to the next level of the competition.

8. Using blocks, build a volume greater than 20 cubes and less than 25 cubes.

9. Using a ruler, sketch a prism that has a volume of 15 cubic centimeters.

10. Using a ruler, sketch a prism that has a volume of 36 cubic centimeters.

11. What is the smallest volume that can be produced using a number cube? Consider whole numbers, not fractions. Technically, you can have a nano-cube, but it would be close to impossible to see and/or work with.

12. Build two different prisms that have 16 cubes each. Sketch your prisms and explain to a neighbor why you think that they have a volume of 16 cubes.

13. You and a friend are playing a game. Your teacher hands you a spinner with numbers 1-6. He asks for both of you to come up with two different prisms each. What are two possible prisms that can be created using the spinner?

1.

2.

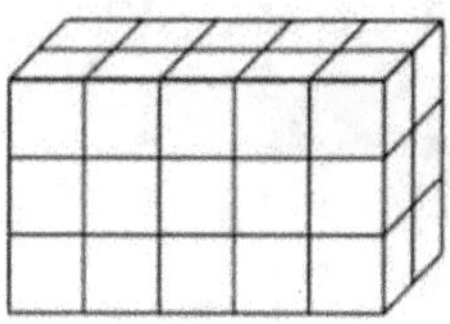

3.

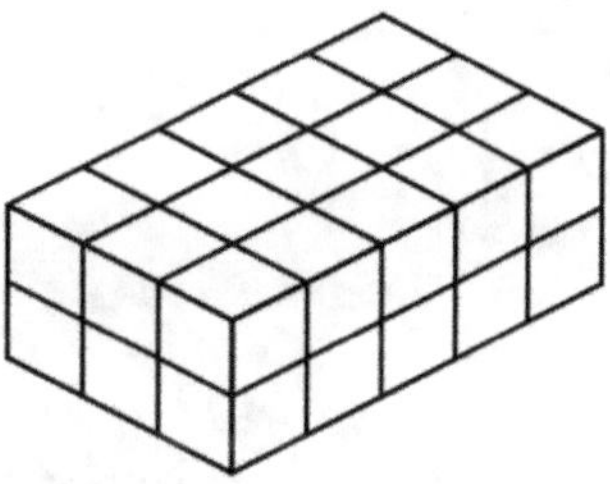

Prism	Number of Cubes in a layer	Number of layers	Number of cubes made to create the prism
1st			
2nd			
3rd			
4th			

Unit 1.2 5.GSR.8.4 Find the Volume of a Rectangular Prism (DOK 1)

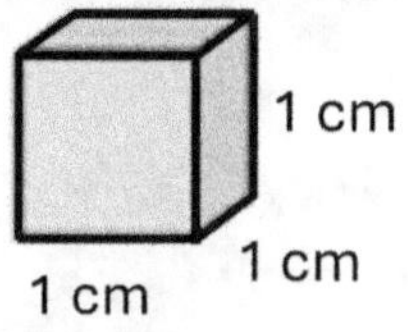

From the last lesson, we learned that finding the volume of a rectangular prism was as simple as taking unit cubes and fitting them so that there were no gaps. Then, we count the number of unit cubes. However, there may be times that you don't have unit cubes with you.

Did you find that you can find the volume of the rectangular prism by multiplying the length, width, and height of the prism?

Formula: Volume = base × height. Volume = Length x Width x Height is another way of expressing the same idea, since to find the area of the base, we multiply the length by the width.

Example 1: Find the volume of a rectangular prism that has a base of 20 square units and a height of 2 units.

Solution: Multiply the base by the height. 20 x 2 = 40 cubic units.

Example 2: Find the volume of a rectangular prism with length = 5, width = 3, and height = 6.

Solution: Multiply length x width x height. 5 x 3 x 6 = 90 cubic units.

Example 3: Jorge wants to find the volume of a rectangular prism with a length of 20 in, a width of 5 in, and a height of 4 in.

Solution: Multiply length x width x height. 20 in x 5 in x 4 in = 400 cubic inches.

Exercises:

1. Find the volume of a rectangular prism with length = 6 in, width = 5 in, and height = 4 in.

2. Find the volume of a rectangular prism with length = 1 cm, width = 3 cm, and height = 7 cm.

3. Find the volume of a rectangular prism with length = 4 ft, width = 6 ft, and height = 10 ft.

4. Find the volume of a rectangular prism with length = 21 in, width = 7 in, and height = 20 in.

5. Find the volume of a rectangular prism with length = 8 mm, width = 10 mm, and height = 3 mm.

6. Find the volume of a rectangular prism with length = 10 mi, width = 12 mi, and height = 5 mi.

7. Find the volume of a rectangular prism with length = 5 cm, width = 10 cm, and height = 15 cm.

8. Conner needs to find the volume of a rectangular prism with length = 6 ft, width = 3 ft, and height = 5 ft.

9. Seth needs to find the volume of a rectangular prism with length = 8 mm, width = 2 mm, and height = 5 mm.

10. Ellen wants to find the volume of a rectangular prism with length = 10 in, width = 50 in, and height = 20 in.

Spiral Quiz

Find the sum.

1. 34+74

2. 86+94

3. 234+23

4. 394+238

Find the difference.

5. 872-238

6. 764-37

7. 1982-1955

8. 1827-238

Find the volume.

9. The base of the rectangular prism is 10 square cm, and the height is 9 cm. What is the volume?

10. The volume of the rectangular prism is 36 cubic units. The height is 3 units. What is the area of the base of the rectangular prism?

Volume: Finding the Volume of Cubes

Example 1: Find the volume of a cube that has side lengths of 3 inches.

Multiply: 3 in × 3 in × 3 in = 27 in^3.

Example 2: Find the volume of a cube that has side lengths of 4 ft.

Multiply: 4 ft × 4 ft × 4 ft = 64 ft^3.

1. Determine the volume of a cube that has side lengths of 8 units.

2. Find the volume of a rectangular prism with side lengths of 7 units.

3. Determine the volume of a cube with side length of 5 cm.

4. Find the volume of a rectangular prism with side lengths of 2 m.

5. Calculate the volume of a rectangular prism that has double the length of a cube with side lengths of 2 m.

6. Determine the volume of a rectangular prism that triples the length of 5 cm.

7. How many unit cubes can be fit into a cube with side length of 12 units?

8. Sketch a cube with a volume of 8 cubic units.

9. Sketch a cube with a volume of 1 cubic unit.

Unit 1.3 5.NR.5.1 Matching Expressions (DOK 2)

Match the following rectangular prisms.

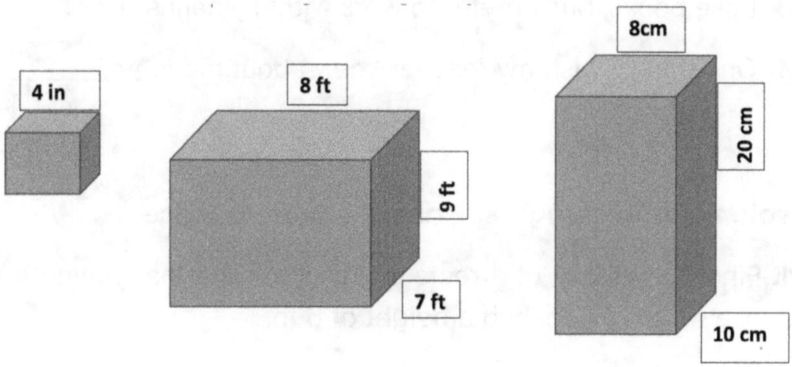

(16) × 4	7 × 72	8 × 10 × 20
8 ×2 + 100	4 + 4 + 4 + 4 + 4 + 44	63 × 8
40 × 40	4^3+0	(4+4) + (4+4)
(3×3)×(4×2) ×7	(4×5)×(2×4)× (5×2)	36 × 2 × 7

Quiz: About Me:

Complete the following questions:

1. I work hard. Answer from 1 to 5. 5 being the most true.

2. I have a job when I get home. T or F?

3. I like books, but I prefer to work with my hands. T or F?

4. One thing I wish my teacher knew about me is ___________

Volume of Rectangular Prisms - More Practice

1. Find the volume of a rectangular prism that has a length of 7 in, a width of 5 in, and a height of 6 in.

2. Calculate the volume of a rectangular prism that has a length of 3 units, a width of 4 units, and a height of 2 units

3. Determine the volume of a rectangular prism that measures 6 units in length, a width of 8 units, and a height of 5 units.

4. Find the volume of a rectangular prism with dimensions 9 units by 2 units by 4 units.

5. Calculate the volume of a rectangular prism that has a length of 7 ft, a width of 5 ft, and a height of 6 ft.

6. Determine the volume of a rectangular prism that has a base of 10 square units and a height of 4 units.

7. Find the volume of a rectangular prism with a base of 40 square units by 3 units.

8. Calculate the volume of a rectangular prism with a length of 10 units, a width of 5 units, and a height of 3 units.

9. Determine the volume of a rectangular prism that has a length of 8 units, a width of 3 units, and a height of 6 units.

Quiz: Find the volume of the following:

1. Maggie has a set of unit cubes. She places 4 in one column. She creates 4 columns with the cubes. She counts the number of cubes. She gets ____ as the number of unit cubes on the square she created. She stacks 2 more layers of the same shape she created. How many cubic units did she use to create the rectangular prism?

2. Greg takes the same number of unit cubes and creates 2 columns with 8 unit cubes. He stacks 2 more layers of the same shape he created. How many unit cubes did he use to create the rectangular prism?

3. Find the volume of a rectangular prism with the base of 20 unit cubes and the height of 4 unit cubes.

4. Find the volume of a rectangular prism with the base of 10 unit cubes and the height of 5 unit cubes.

5. The expression 3(4+4+4) is the volume of a rectangular prism. Is the expression 5(20-16) equivalent to the previous expression? Yes or No? Explain your work.

Multiplication and Division Equations

Example 1: $2x = 10$ This is a multiplication equation, so we use the opposite operation in order to solve for the variable.

Ans. $x = 5$

Example 2: $x/3 = 15$ This is a division equation, so we use the opposite operation in order to solve for the variable.

Ans. $x = 45$

Solve for the variable.

1. $3x = 60$

2. $x/4 = 9$

3. $x/5 = 30$

4. $4x = 90$

5. $10x = 150$

6. $x/7 = 8$

7. $9x = 72$

8. $11x = 121$

9. $x/6 = 40$

10. $30x = 1{,}800$

Unit 1.4 5.NR.5.1 Understanding Numerical Expressions PEMDAS

Numerical expressions are a set of numbers and operations.

Example 1: 4+7×6

Example 2: 3+3×7-3

Example 3: 10-8+3×4

Let's Consider The Order of Operations. The concept of the order of operations using PEMDAS (Parentheses, Exponents, Multiplication, and Division (from left to right), Addition and Subtraction (from left to right). It is a way to remember which steps to take in order to simplify a numerical expression correctly.

Example 4: Evaluate 5 + 2 × 9. First, multiply 9 and 2, then add 5. Then, 5 + 18 = 23

Example 5: Expression with Grouping Symbols

Grouping symbols, such as parentheses and brackets, are used to group parts of an expression that should be evaluated first.

Example: (6-2) × 7. Subtract 6 and 2 first, then multiply by 7.

Evaluate: (2 x 10) + (3 x 5). (20) + (15) = 35

Example 6: Evaluate (6-2) x 7 (4) x 7 = 28.

Challenge Problem. Given the following numbers:

1 2 3 4 5 6 7 8 9 10

Use grouping symbols along with addition, subtraction, multiplication, and division to come up with the simplified solution of 100 in three different ways.

Real-Life Application

Content: Show how numerical expressions can represent real-life situations. For instance, calculating the total cost of items bought.

Example: If a hotdog costs $3 and a cola costs $2, then buying 2 hotdogs and 3 colas can be represented as 2 × 3 + 3 × 2.

Activity: Create two real-life scenarios and write corresponding numerical expressions.

Extended Activity: Create a foldable in order to write a 5-page short story with numerical expressions. Be prepared to share with your group and/or class.

Unit 1 Review

Standards Covered:

5.GSR.8.3 Investigate the volume of right rectangular prisms by packing them with unit cubes without gaps or overlaps. Then, determine the total volume to solve problems. 5.GSR.8.4 Discover and explain how the volume of a right rectangular prism can be found by multiplying the area of the base times the height to solve real-life, mathematical problems. 5.NR.5.1 Write, interpret, and evaluate simple numerical expressions involving whole numbers with or without grouping symbols to represent real-life situations.

Unit cubes are packed into the bottom of rectangular prisms without gaps or overlaps. Find the number of unit cubes in the rectangular prisms.

1. The area of the base is 50, and there are 4 more layers added on top of the bottom layer. How many unit cubes were used to pack into the rectangular prism? A. 100 B. 150 C. 200 D. 250

2. The area of the base is 30, and an additional layer was added on top of the bottom layer. How many unit cubes were used to pack into the rectangular prism? What is the total volume? A. 10 B. 20 C. 30 D. 60

3. Josh measured the length of the base of the rectangular prism to be 9 unit cubes. He counted the width of the base to have 8 unit cubes. Once Scott had filled in the base of the rectangular prism, he added 2 more layers of unit cubes. How many unit cubes are in the rectangular prism? A. 132 B. 72 C. 164 D. 216

4. Conner used 80 unit cubes to create a rectangular prism. If he had 4 layers of unit cubes, what could be the area of the base of the rectangular prism? A. 10 B. 15 C. 20 D. 25

5. What is the volume of the rectangular prism with base 12 and a height of 1 cm?

A. 10 B. 12 C. 14 D. 1

6. What is the volume of the rectangular prism in #5 if there were 3 layers like that added to the rectangular prism?

A. 20 B. 36 C. 48

7. What is the volume of the rectangular prism with base 12 and a height of 1 cm?

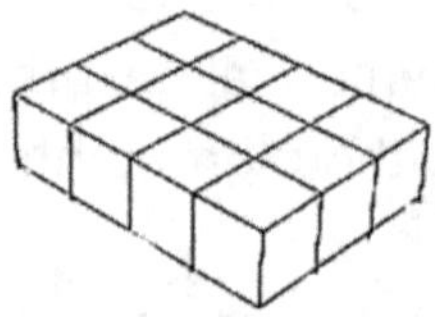

A. 10 B. 40 C. 50 D. 60

8. If the volume of the rectangular prism in #7 is tripled, what would be the volume of the new rectangular prism?

A. 50 B. 120 C. 180 D. 200

Consider the following numeric expressions: Write, interpret, and evaluate them.

9. The sum of 24 and 48

10. The product of 16 and 39.

11. The number 4 plus the sum of 21 and 19.

12. The difference of 62 and 9.

13. 15(2 +8)

14. 5(7) - 4(5)

15. 10×32 + 7×8

Unit Plan: Unit 2

Unit 2 Vocabulary

These are the vocabulary words to be discussed in this unit. Try to have fun with them and discuss them with your learning partners and stakeholders, i.e., parents, students, coaches.

Conversion: To change from one unit of measure to another.

Length: How long something is.

Line Plot: Points on a line.

Liquid Volume: The amount of space that liquid takes up.

Mass: The quantity of matter in an object.

Measurement: Process of assigning a number value.

Metric and Customary Systems: Systems used to measure objects.

Metric and customary units of measure: Ways to measure.

Numerical Patterns: A sequence of numbers or objects.

Power of ten: 10 multiplied by itself a certain number of times.

Rules: A set of guidelines that defines a relationship.

Weight: How heavy an object is.

Create a word search with the words. Share among friends.

Unit 2 5.NR.1 Place Value (DOK 1)

The powers of 10 are used to describe larger numbers. For example, the number $100 = 10 \times 10 = 10^2$. The number $1{,}000 = 10\times10\times10 = 10^3$.

Example 1: Have you ever wondered what the distance from the Earth to the moon is? 238,855 miles. We could round it to 240,000 miles just to get an idea of place value. We could say that $240{,}000 = 24 \times 10^4$ miles. The distance from Earth to the sun is 93 million miles. How could you write this as a power of 10? 93×10^6

Example 2: The distance from Georgia to California is 2100 miles. How could this number be written as a power of 10? 21×10^2

Example 3: Is 450 divided by 10 the same as multiplying 450 by 1/10?

$450 \div 10 = 45$

$450/10 = 45$

$(450 \text{ x } 1/10) = 45$

Ans. Yes

10^4	10^3	10^2	10	10^0
Ten thousands	Thousands	Hundreds	Tens	Ones

Example 4: Multiply: $4.5 \text{ x } 10^3$ (use an exponential 3)
Ans. 4,500

Write the following numbers as a power of 10.

1. 7,500

2. 830,000

3. 12,000

4. 490

5. 680,000

Write the following numbers in standard form.

6. 4.5×10^3

7. 3.7×10^2

8. 63×10^4

9. 81×10^5

10. 9.6×10^3

11. 3.8×10^5

12. 12.42×10^4

13. 7.3×10^3

14. 52×10^2

15. 99×10^3

Spiral Quiz

1. Add 1477+289

2. Subtract 1873 - 368

3. Multiply 356 × 12

4. Divide 289/17

5. Simplify 2(3+4)÷7

6. 36 in = ____ ft

7. 5280 ft = _____ mile(s)

Unit 2.1 5.NR.1.1 Use Place Value (DOK 1)

Chris and Kaynan are at the mall with their parents. Their mom gave them both $20 and said to have fun. Chris had done four chores to earn his money. Each chore took Chris about half an hour to complete. Kaynan had babysat the neighbor's kid for two hours. Who earned the most money per hour?

While shopping, Kaynan bought a candy bar for $1. Chris told Kaynan that he had more money than him because his change was one $10, $5, and four $1 bills. Is he correct? Why or why not?

Answer: $19 is less than $20, even if it is just a $20 bill. (We've all had this game played on us at one time or another.) Considering money and digits, we can picture the ones, tens, hundreds, thousands, and ten thousands.

Ten thousands	Thousands	Hundreds	Tens	Ones

Consider the number 23,567: The 2 is in the ten-thousands, the 3 is in the thousands, the 5 is in the hundreds, the 6 is in the tens, and the 7 is in the ones place.

Ex. 1 Given the number 75,638, what digit is the 5?

Ans. The 5 is in the thousands place.

Ex. 2 Given the number 78,634, how many times is the 8 from the 4?

Ans. The 8 is 10^3 from the 4 place.

1. Given the number 964, show it in the place-value chart.

2. What is the value of 4, given the number 98,435?

3. What is the value of 8, given the number of 37,284?

4. What is the value of a digit in the hundreds place, given the number 938,792?

5. What is the value of the digit in the ten-thousands place, given the number 23,748?

6. Show the number of 865 written in expanded form.

7. Show the number of 8,761 written in expanded form.

8. Given the expanded form of the following number: 20,000 + 7,000 + 600 + 50 + 2, write the number in standard form.

9. Given the expanded form of the following number: 30,000 + 8,000 + 500 + 40 + 3, write the number in standard form.

10. How many times is the value of 8 from the number 5, given the number 8,576?

Given the following numbers, what is the place value of the number 8?

11. 82,345

12. 68,392

13. 19,820

14. 27,348

15. 27,829

Given the following numbers, how many times is the number 8 greater than the number 5?

16. 850

17. 4,805

18. 5,842

19. 6,835

20. 758

21. True or False? 400 is ten times greater than 4.

22. True or False? 300 is times greater than 30.

Unit 2.2 5.NR.1.1 Cont.

Ex. 1 Given the following numbers: 0.5, 5, 50. How are they related?

Ans. 0.5 is 1/10 times 5. The number 5 is 1/10 of 50. The number 5 is 10 times greater than 0.5.

Ex. 2 Give the following numbers: 0.8 and 8. How are they alike?

Ans. Alike: Both contain the number 8. Different: 0.8 is 1/10 times than 8.

1. Given the following numbers, compare and contrast the numbers 0.07 and 0.7.

2. Given the following numbers, compare and contrast the numbers 20 and 0.2.

3, Given the following numbers: 0.4, 4, and 40. How are they related?

4. Given the following numbers: 0.03, 0.3, and 3. How are they related?

Ex. 3 A bag of candy contains 0.88 grams of candy. There are 8 pieces of candy in the bag. How much does one piece of candy weigh?

Ans. One piece of candy weighs 0.11 grams.

5. A box of paper clips weighs 1.44 grams. There are 12 paper clips in the box. How much does one paper clip weigh?

6. A box of wafers weighs 2.4 grams. There are 12 wafers in the box. How much does one wafer weigh?

Ex. 4 You have a friend who is having a hard time understanding what base ten blocks are used for in Math. Your teacher asked you to help explain to your friend how the number 3.5 is represented using Base 10 blocks.

Ans. You draw 3 whole base, 10 rods, and one half of another.

7. Represent the number 2.8 using Base 10 blocks.

8. Represent the number 5.2 using Base 10 blocks.

9. Represent the number 2.8 + 5.2 using Base 10 blocks.

10. Represent the number 5.2 - 2.8 using Base 10 blocks.

Unit 2.2 5.NR.1.2 Patterns in Powers of Ten (DOK 2)

When working with numbers, sometimes it is easier to work with numbers in the Powers of Ten. Hiking through the North Georgia mountains, one may come upon Burnt Mountain near Jasper, Georgia. The summit of Burnt Mountain is 3,200 feet above sea level, which can be written as 32×10^2. These are approximations to show variations and relevance in height.

1. Fort Mountain in Chatsworth, GA, is 285×10. Write the number in standard form.
2. Blood Mountain, bordering Lumpkin and Union Counties, has an approximate height of 4,500 feet above sea level. What would the number be in Powers of Ten?
3. Brasstown Bald, bordering Towns and Union counties, has an approximate height of Blood Mountain plus 3×10^2 feet. What is the approximate height of Brasstown Bald?
4. Kuwohi Mountain in North Carolina is almost 38×10^2 feet higher than Fort Mountain. What is the height of Kuwohi Mountain?
5. Black Elk Peak in South Dakota is almost 10×10^3 feet lower than $10 \times$ feet. What is the height of Black Elk Peak?

Equation	Yes or No	Justification
$10 \times 10 \times 10 = 10^3$		
$36 \times 10^2 = 360$		
$10^3 = 30$		
$289 \div 10^2 = 28.9$		
$3.75 \times 10^3 = 3750$		
$145.5 \div 10^3 = 0.1455$		

Addition and Subtraction Equations

Example 1: $x + 12 = 36$ This is an addition equation. Use the opposite operation in order to solve for the variable. We will subtract 12 from both sides. Ans. $x = 24$.

1. $x + 9 = 27$

2. $x - 8 = 24$

3. $x - 7 = 21$

4. $x + 10 = 42$

5. $x - 15 = 45$

6. $x + 11 = 48$

7. $x - 12 = 36$

8. $x + 9 = 81$

9. $x + 18 = 36$

10. $x - 25 = 75$

Unit 2.3 5.MDR.7.1 Units of Measurement (DOK 2)

Every day, we use various units of measurement to make the world meaningful around us. Some of the units of measurement are distance, weight, volume, and time. Distance is the amount of space between two points. It can be measured using units like kilometers, meters, centimeters, etc.

Which unit would be the most reasonable measure?

1. Width of a car: 2 cm, 2 m, 2 km

2. Weight of a deer: 300 mg, 200 g, 300 kg

Example 1: Determine how many centimeters are in a meter.

Ans: 1 meter (m) = 100 centimeters (cm). We can reference the metric table to check our work. You may remember that there are 100 years in a century, so it would be likely that there are 100 centimeters in a meter.

Weight measures the heaviness of an object. Common units include kilograms, grams, pounds, tons, etc.

Example 2: Determine the amount of grams in a kilogram.

Ans: 1 kilogram (kg) = 1000 grams (g). Again, we can reference our metric chart to check our work.

Volume is the amount of space occupied by an object. Units used for measuring volume include liters, milliliters, cubic meters, etc.

Example 3: Determine the amount of milliliters in a liter.

Ans: 1 liter (L) = 1000 milliliters (ml)

Time is a measure of the duration of an event. It is typically measured in units like seconds, minutes, hours, days, etc.

Example 4: If we know that there are 60 seconds in a minute, how many seconds are in an hour?

Ans: 1 minute (min) = 60 seconds (s), and 1 hour (hr) = 60 minutes (min). Therefore, we can say that there are 3600 seconds in an hour.

Try your hand at conversions:

1. Change 7.5 m to centimeters.

2. How many grams are in 2.4 kilograms?

3. How many milliliters are in 5.5 liters?

4. If an ATV travels 30 miles per hour, how far will it go in 4 hours?

5. Convert 42 hours to seconds.

6. 5 pounds in kilograms?

7. How many liters are in 1,500 milliliters?

8. How many centimeters are in 2.1 meters?

9. Convert 4.75 liters to cubic meters.

10. If a movie is 1.5 hours long, how many minutes is it?

11. How many kilometers are in 0.6 miles?

12. How many meters is 90,000 millimeters?

Your answers should appear in this box.

7	2	1	7	8	0
5	5	0	0	1	9
0	2	2	1	0	0
0	0	9	2	9	7
2	4	0	0	6	4
0	0	4	7	6	8

Unit 2.4 5.MDR. 7.2 Asking Questions and Using Graphical Displays to Solve Everyday Problems (DOK 2)

Today, we will learn how to ask questions, gather information, and use graphs to solve everyday problems. Graphs help us understand data visually, making it easier to find solutions to various situations we encounter in our daily lives.

Example 1: Students are measuring the length of toys on a desk to the nearest 1/8, 1/4, and 1/2 of an inch. How many toys have a length of 1/4 in? How many toys measure 1/2 in? What would the measure be if the lengths of all the toys are added together? Ans. 3 toys measure 1/4 in, and 1 measure 1/2 in. Ans. 3 1/2 in.

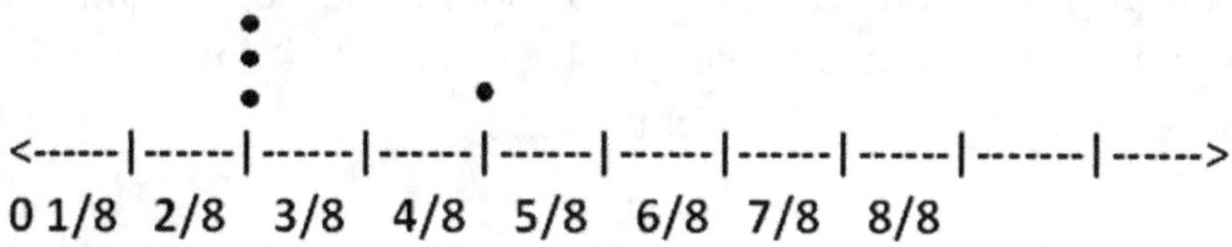

Example 2: Making a Weekly Schedule Katie wants to plan her week to balance her time between homework, chores, and exercise. She creates a schedule to help her track her time. How can Katie create a pie chart to represent her weekly schedule and ensure she allocates enough time to each activity?

Gather Information: Katie tracks her activities for a week and tracks data on the time spent on each task. She creates a pie chart showing the percentage of time spent on homework, chores, and hobbies. She multiplies the percentage by the number of degrees of a circle (360 degrees). This will give the corresponding angle of the sectors.

Kirk receives a monthly paycheck and wants to track how he spends his money on home, transportation, and utilities. How can Kirk use a bar graph to compare his expenses on home, transportation, and utilities over a month?

Gathering Information: Kirk records his spending on home, transportation, and utilities for a month and notes down the amounts. He creates a bar graph showing the amount spent on each category for the month.

1. Your class is studying weather patterns. Your group wants to analyze the tornado track variations over a ten-year period. What would be a good graphical display to show that there may be more than one "tornado alley" in a state?

2. Your family complains that the gas prices are horrible. You want to take a break and go to the beach. What type of data can you gather to validate your argument that prices decline during the summer months?

3. You want to surprise your sister with something for her birthday. You only make $200 a month landscaping your neighbor's lawn. You think about it and decide to create a budget. What are some things that you may include in the budget?

Data Data

Data is everywhere, it seems. Somebody is collecting something and/or studying something. Take a minute to think of some examples of data that are collected. Grades, income, etc. In baseball, the number of home runs, the length of a home run, etc. Data is collected and studied meticulously and in depth. Consider a bank or the Stock Exchange, data is collected to show trends in order to make specific goals or predictions about the future. Let's look at a few examples. Let's start by finding the mean of certain samples. Mean is another word for average.

6.NR.2.1 Describe and interpret the center of the distribution by the equal share value (mean).

Example 1: Freddy the Hall of Fame pitcher is having another banner season. Tonight is his debut against the Lake City Mockingbirds. There is game one and two played tonight, and a third game if a tie. Freddy does a good job in game 1, closing out 8 batters. In game 2, his team lost by a run. He closed out 5 and walked 3. In game 3, he injures his shoulder and can only get 2 players at bat to strike out. What is the mean (average) of the number of batters he struck out in the three games? We'd add 8 + 5 + 2 = 15. Next, we'd divide by the number of games played against an opponent. Therefore, 15/3 = 5.

Example 2: Myra plays basketball for her middle school. She's had a good season. She wants to tie the record for scoring the most during the 12 games she's played. The record was 16 points per game. She has scored an average of 14 in the first ten games. How many points does she need to score to break the record? ((14*10) + 2x)/13 = 13.

Step 1: Create an equation or expression to solve for x. $\frac{14 \times 10 + 2x}{12} = 13.$

Step 2: Simplify. 140 + 2x = 156 by subtracting 140 from both sides.

Step 3: Solve for x: 2x=16 by dividing by 2. So, x = 8.

Find the measure of center, mean.

1. Test scores: 87, 83, 90, 95, 100, 88

2. Football scores: 35, 28, 24, 16, 18

3. Soccer scores: 5, 4, 3, 2, 6, 4

4. Miles per gallon: 40.2, 36.4, 34.7, 36.5, 38.8, 39.4

5. Miles per hour: 55, 65, 70, 63, 68, 67

6. Chicks born today: 150, 254, 156, 143, 157

NR.2.2 Summarize categorical and quantitative (numerical) data sets in relation to the context: display the distributions of quantitative (numerical) data in plots on a number line, including dot plots, histograms, and box plots and display the distribution of categorical data using bar graphs.

Example 1: Dot Plot

In creating a dot plot, we draw a line that includes the range of scores. We'll place a dot (or x) above each score we use on the number line.

If we keep the previous soccer scores: {5, 4, 3, 2, 6, 4}. We can use them to plot these scores on the line.

```
                 x
         x   x   x   x   x
0---1---2---3---4---5---6---7---8---9---10
```

Once the scores are plotted on a number line, we can make observations and generalizations about the scores. How many scores were greater than or equal to 4?

$x \geq 4$? {4,4,5,6}. Four scores were greater than or equal to 4.

Example 2: A class wants to examine Math test scores on an exam to see if it was a fair test. The scores are 80, 95, 95, 80, 85, 100, 95, 95, 80, 90, 80, 80.

```
               X
               X                   X
               X                   X
               X                   X
               X     X     X       X     X
70 --- 75 --- 80 --- 85 --- 90 --- 95 --- 100
```

Students scored 80 or above.

Exercise:

Create the dot plots with the following information.

1. Test Scores 70, 75, 80, 75, 80, 75, 85, 90, 80, 90

2. Number of Fish Caught: 5, 3, 5, 4, 5, 5, 0 , 1, 2, 3, 1, 3, 5

3. Football Scores: 7, 10, 7, 14, 7, 21, 28

4. Miles ran: 4.5, 2.5, 3, 2, 4, 5, 4.5, 4, 3, 2.5, 3, 3, 1.5

5. Laps in the pool: 2 1/2, 3, 2, 2 1/2, 1 1/2, 2 1/2, 1, 2, 2 1/2

A histogram is similar to a bar graph. It is used for quantitative data. The data is separated into intervals.

Example: The students at Mountain Bluff Elementary were given a survey to study the number of hours they spent on a certain website away from school.

35, 34, 43, 19, 58, 15, 60, 18, 38, 25, 17, 51, 48, 18, 23, 21, 14, 44, 34, 39, 28, 27

Step 1: Determine the number of points by counting the number of points in the list. (n=20)

Step 2: Compute the range of the data, R. (R = 60 - 14 = 46)

Step 3: Determine the number of classes, K. $(k = \sqrt{n}, k = \sqrt{22} \approx 4.69)$

Step 4: Compute class width, W=R/K. (w = 46/4.69) 4.69 can be rounded up to 5. 46/5 = 9.2
Rounding up 9.2 to 10.

Step 5: Determine the starting points of the class intervals

Class 1: 10 to less than 20 (10-19)

Class 2: 20 to less than 30 (20-29)

Class 3: 30 to less than 40 (30-39)

Class 4: 40 to less than 50 (40-49)

Class 5: 50 to less than 60 (50-59)

Class 6: 60 to less than 70 (60-69)

We need 6 classes since 6 falls on the boundary of the maximum class.

Step 6: Determine the starting points of the class intervals.

Step 7: Use a tally sheet to keep track of the data.

	Data Points	Tally	Frequency
10 to < 20	19, 15, 18, 17, 18, 14	HI	6
20 to < 30	25, 23, 21, 28, 27	HI	5
30 to < 40	35, 34, 38, 34, 39	HI	5
40 to < 50	43, 48, 44	III	3
50 to < 60	58, 51	II	2
60 to < 70	60	I	1
Total			22

Step 8: Construct the Frequency Distribution Table and/or Histogram.

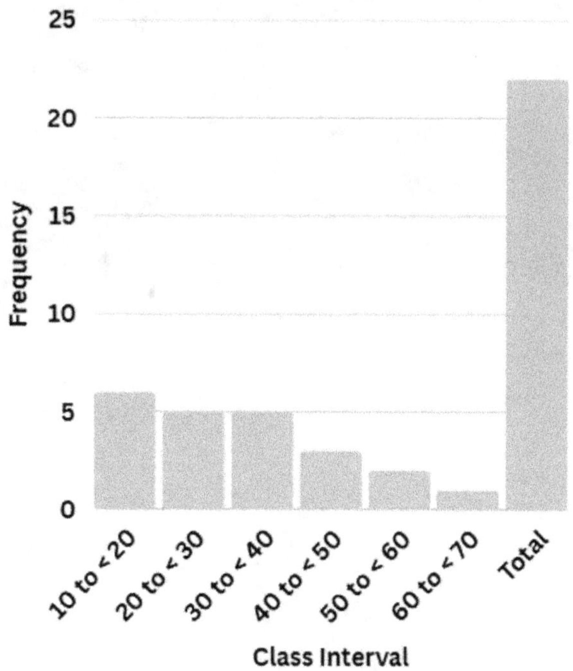

Unit 2.5 5.MDR.7.3 Conversions in the Metric System (DOK 2)

When you convert among units in the metric system, you change the size of a measurement from one unit to another. For example, if you need to change from grams to kilograms to understand how much something weighs.

Kilo 1,000	Hecto 100	Deca 10	Unit 1	Deci 0.1	Centi 0.01	Milli 0.001
Kilogram	Hectogram	Decagram	Grams	Decigram	Centigram	Milligram
Kiloliter	Hectoliter	Hectogram	Liters	Deciliter	Centiliter	Milliliter
Kilometer	Hectometer	Decameter	Meters	Decimeter	Centimeter	Millimeter

Understanding the Metric System:

The metric system is like a set of building blocks for measuring things.

The metric system uses units like grams, liters, and meters. It is easier to convert between units since they are based on powers of 10. In order to convert units, you need to know how many of one unit equals another. For example, 1 meter is equal to 100 centimeters. You can use conversion factors or multiplication and division to change from one unit to another.

Example 1: There are 500 milliliters of water in the swimming pool. Sheri wants to know how many liters that is since she wants to go swimming for her birthday. How can she determine how many liters are in the pool? Ans: She could divide by 1,000 since 1 liter is equal to 1,000 milliliters.

Did you catch the size of 500 milliliters? Lol. It is a small matter. A standard swimming pool in a backyard has about 50,000 liters. If we changed that to ml, then it would be quite a few ml. 50,000,000 ml

Example 2: You are baking peach tarts for the weekend fair. Your sister tells you that there are 1,500 grams of peaches, but your recipe requires kilograms. How would you explain to her that you have enough peaches? Ans: You would divide by 1,000 since 1 kilogram is equal to 1,000 grams.

Example 3: You have been voluntold to help decorate a float. You need to measure the length of a streamer in centimeters; however, your ruler is divided into millimeters. How can you measure the streamer with your ruler to determine how many centimeters your float requires? Ans: You need to convert millimeters to centimeters by dividing by 10.

Convert:

750 grams =________kilograms	4500 = 9	0
800 milliliters = ______________ liters	2900 = 5	S
4.5 kilograms = ______________ grams	87 =	I
6.5 liters = ______________ milliliters	6500 = 15	L
360 centimeters = ______________ meters	3.6 = 25	Y
29 kilometers = ______________ decameters	54500 =	E
545 meters = ______________ centimeters	.75 = 9	V
870 hectograms = ______ grams	8 = 13	u

Use the key to decode the phrase!

9.12.15.22.5.25.15.21

5.MDR.7.4 Convert among units within relative sizes of measurement units within the customary measurement system. (DOK 2)

Unit 2 – Dog Show

Lucky you, you were invited by a friend to a local dog show. You've seen a couple of dog shows on television in the past, and think, "why not?" Upon arrival to the event, you noticed that the show was humming with excitement. You saw Great Danes, German Shepherds, even Tea Cup Chihuahuas.

The first event was a weigh-in for the dog breeds. Which measurement do you think would be used in measuring the weight of the dogs? Grams, milligrams, or kilograms? Discuss with a friend.

Dog Breed	Kilograms	Decigrams	Milligrams
Great Dane	60		
German Shepherd		300	
Pitbull			3000
Tea Cup Chihuahua	2		

Another event is the races. To be fair, the dogs were classified in large, medium, and small breed dogs. A mutual friend, Scott, has entered his Pittie, Scooter, into the race. Scooter can be persuaded with an occasional snack, but he is more concerned about a nap than a race.

Dog Breed Race

Dog Breed Wins	Meters	Centimeters	Millimeters
Great Dane	500		
Labrador		3000	
Pitbull			10

Are there any patterns that you noticed, other than Scooter having rolled over during his nap?

During the agility part of the program, dogs were thrown frisbees to see which breed would have the most graceful, highest leap, etc.

With a partner, discuss and create a new activity for the dogs. How would you decide who won the contest? In order to take the title, "Talented Pup" home." Be creative and willing to share if needed.

Unit 2.6 5.MDR.7.4 Conversion Factors

5.MDR.7.4 Convert among units within relative sizes of measurement units within the customary measurement system.

12 inches	1 foot
36 inches	3 foot
1 mile	5280 feet
3 feet	1 yard
1 cup	16 ounces
4 quarts	1 gallon

Consider converting 12 feet to yards. Since 1 yard is 3 feet, then 12 feet is 4 yards.

Let's consider which units of measures to use in order to get an idea about which measurements to use.

1. The length of a room

2. The height of a table

3. The length of a pencil

4. The width of your arm

5. The length of your arm

Tell which measurement is needed.

6. Finding the length and width of a poster

7. Finding the distance between Chattanooga, TN and Miami, FL.

8. Finding the height of a 30-story building

9. Finding the length of a table

10. Finding the width of a sheet of notebook paper

Example 1: Convert 5 miles to yards. (1 mile = 5,280 ft) (1 yard = 3 ft)

Convert:

1. 4 miles = ____________ yards

2. 30 inches = ___________ feet

3. 50 pounds = ____________ ounces

4. 1.5 miles = ____________ inches

5. 32 ounces = ___________ pounds

6. 2 gallons = ____________ cups

7. 2 miles = _____________ yards

8. 1.25 tons = ____________ pounds

9. 10 degrees Fahrenheit = ________ Celsius

10. 8 quarts = _____________ gallons

11. 28 cups = _____________ gallons

12. 50 yards = _________ feet

13. 10 pints = ____________ cups

14. 40 pounds = __________ tons

15. 64 cups = ____________ gallons

16. 800 inches = __________ feet

17. How many ounces are in a ton?

18. How many inches are in a mile?

19. How do you convert square feet to square miles?

20. How do you convert square inches to square feet?

Compare. Determine >, <, or = ___.

21. 3 ft ___ 38 in

22. 4 yd 2 ft ___ 15 ft

23. 5 mi ___ 36,000 ft

24. 150 in ___ 13 ft

25. 6 yd 2 ft ___ 420 in

Which unit should be used to measure?

26. The height of a camel

27. The length of a crocodile

28. The length of a car

29. The width of a football field

30. The distance a plane flies

Solve.

31. A dog is 3 ft long. A kitten is 8 in long. How many feet longer is the dog compared to the kitten?

32. Explain how to change 14 yds to feet and inches.

33. How many inches is in three miles?

Which unit would you use to measure the following? Consider lbs., oz., T, fl oz., gal., cup

34. A doe weighs about 120 ___.

35. A residential pool hold about 12,000 ___ of water.

36. A small serving of cereal is 8 ___.

37. A truck weighs about 2 ___.

Unit 2: Conversion of Time.

The conversion of time typically involves converting between different units of time, such as seconds, minutes, hours, days, weeks, and months.

1 minute	60 seconds
1 hour	60 minutes
1 day	24 hours
1 week	7 days

Example 1: Covert 240 minutes to hours

Ans. To convert 240 minutes to hours, divide 240 minutes by 60 minutes.

Example 2: Convert 4 hours to minutes.

Answer: multiply by 60 since there are 60 minutes in an hour. Therefore, 4 hours = 240 minutes.

Example 3: Convert 21 days to weeks.

Ans. To convert 21 days to weeks, divide 21 days by 7. Therefore, 21 days = 3 weeks.

1. Convert 180 seconds to minutes.

2. Convert 6 hours to minutes.

3. Convert 20 days to hours.

4. Convert 360 minutes to hours.

5. Convert 35 weeks to days.

6. Convert 7 hours to seconds.

7. Convert 30 minutes to hours.

8. Convert 2,400 seconds to minutes.

9. Convert 960 minutes to hours.

10. Convert 9 months to years.

11. Convert 490 days to weeks.

12. Convert 15 hours to days.

13. Convert 240 seconds to minutes.

14. Convert 12 weeks to days.

15. Convert 4,800 seconds to minutes.

Match:

7	2400	12	30	180	6	960	490	20	35	9	15	240	4800
420	40	84	.5	10,800	360	16	70	480	245	.75	0.625	4	60

Unit 2 Review

5 NR.1.1 Explain that in a multi-digit number, a digit in one place represents 10 times as much as it represents in the place to its right, and 1/10 of what it represents in the place to its left. 5.NR.1.2 Explain patterns in the placement of digits when multiplied or divided by a power of 10. Use whole number exponents to denote powers of 10, up to 103. 5.MDR.7.3 Convert among units within the metric system and then apply these conversions to solve multi-step, real life problems. 5.MDR.7.4 Convert among units within relative sizes of measurement units within the customary measurement system. 5.MDR.7.1 Explore real-life problems involving different units of measurement, including distance, weight, volume, and time. 5.MDR.7.2 Ask questions and answer them based on gathered information, observations, and appropriate graphical displays to solve problems relevant to everyday life.

In a multi-digit number, a digit in one place represents 10 times as much as it represents in the place on its right. It is 1/10 of what it represents in the place to its left.

1. Consider the hundreds place and the tens place. Compare the hundreds place to the tens place.

2. Consider the thousands place and the hundreds place. Compare the thousands place to the hundreds place.

3. The value of the 5 in the number 86,532 is ____ times the value of the 5 in the number 72,853.

4. The value of the 3 in the number 93,871 is ___ times the value of the 3 in the number 30,765.

5. Write the following number in standard form. 80,000 + 7,000 + 300 + 20 + 1

6. Write the following number in expanded form. 74,293

Powers of 10.

7. Multiply 754 x 100.

8. Multiply 82 x 10,000.

9. When multiplying 76 by 1,000, how many powers of 10 is 76 multiplied by?

10. Multiply 53 x 10^4.

11. Divide 86,000 by 10^3.

12. Divide 6,500,000 by 10^4.

Write the following numbers as a factor times the power of 10.

13. 831,000,000

14. 670,000

15. 920,000,000

Convert the following within the metric system.

16. 18 km = _____ m

17. 23 m = _____ cm

18. 176 dm = ____ m

19. 240 g = ____ kg

Convert the following within the customary system.

20. 3 miles = ____ ft

21. 24 ft = ____ yd

22. 19 in = ____ ft

Answer the following questions:

23. A rectangle has a length of 15 in and a width of 10 in. What is the area of the rectangle? What is the perimeter of the rectangle?

24. If one box contains 16 crayons, how many crayons are in a carton of 60 boxes?

25. A baker wants to sell brownies for the festival. If he bakes 60 brownies, how many dozen brownies can he sell?

26. Callie makes $430 per week. Her brother makes $1,340 per week. What is the place value for the 4 in each number?

27. Gina makes two trips that take her 3 hours to complete for her job. If she decides to take one more trip each day, how many more hours each week will she be spending on the road?

Unit 3:

Building Conceptual Understanding of Multiplication and Division with Whole Numbers

Unit 3 Vocabulary: Building Conceptual Understanding of Multiplication and Division with Whole Numbers

These are the vocabulary words to be discussed in this unit. Try to have fun with them and discuss them with your learning partners and stakeholders, i.e., parents, students, coaches.

Algorithm: A well-defined step-by-step procedure used to solve certain math problems.

Distributive Property: Multiplying a number by a sum of numbers equals multiplying a number by each number separately and then adding them.

Dividend: The number that is being divided in a division problem.

Divisor: The number by which another number, the dividend, is divided.

Equation: A statement with two equal expressions connected by an equals sign.

Expression: A statement that uses numbers, operations, and variables to represent things in the real world.

Measurement Division (repeated subtraction): A division process of dividing a quantity by equal groups.

Multiplicand: The number that is being multiplied by another number.

Multiplier: The number that is used to multiply a number.

Order of Operations: A set of rules used to simplify expressions.

Partition Division (or fair-sharing): A method of breaking larger numbers into smaller numbers.

Partial Product: A method of multiplying a part of a number by a part of another number.

Partial Quotient: A method of dividing a part of a number by a part of another number.

Product: Another word for multiplication.

Properties of Operations: Rules that govern how math operations behave.

Quotient: Result from dividing two numbers.

Remainder: An amount, or number, leftover from dividing another number.

Pick half the definitions to read for 5 minutes, and discuss with a partner who has read the other half of the definitions.

5.NR.2.1 Fluently multiply multi-digit (up to 3-digit by 2-digit) whole numbers to solve real-life problems. (DOK 2)

Show your work by creating an area model and multiplying the numbers. Check your work for accuracy.

Example 1

```
  54
 ×12
 108
 54
 648
```

Step 1: multiply 2 by 54.

Step 2: multiply 1 by 54 and move over one digit space to the left.

Step 3: add.

Therefore, 54 × 12 = 648

Part 2: Multiply 54 x 12 using Area Models.

Example 2

Show your work.

1. 75x20	6. 271x33
2. 81x93	7. 687x29
3. 34x72	8. 159x82

4. 25x82	9. 724x30
5. 39x28	10. 389x74

Order of Operations

A way to simplify expressions is called the Order of Operations. This set of rules gives a step by step way of getting the correct answer. Without using the correct steps in the order of operations, one can get an incorrect answer.

Please Excuse My Dear Aunt Sally is a mnemonic devise to help you remember which steps to take along the way.

Please : Parentheses

Excuse: Exponents

My: Multiplication Left to Right

Dear: Divide Left to Right

Aunt: Add Left to Right

Sally: Subtract Left to Right

Let's look at a couple of examples.

Ex. 1 $(24 \div 2)^2 \times (2 \times 5) - (6-3) \div 3$

Step 1: Simplify $(12)^2 \times 10 - 3 \div 3$

Step 2: Simplify $144 \times 10 - 3 \div 3$

Step 3: Continue to simplify $1440 - 1 = 1439$

Ex 2 $(64 \div 8)^2 - (3-1)^3$

Step 1: Simplify $(8)^2 - (2)^3$

Step 2: Simplify $64 - 8 = 56$

Simplify.

1. $8 + (7 - 2)$

2. $(5 + 4) + (9 - 4)$

3. $(9 - 3) \times 5$

4. $20 + (5 + 8) - 10$

5. $(10 + 12) \times (6 \div 3)$

6. $(8 - 2) - 2$

7. $8 + (21 \div 7) \times 2$

8. $(10 \times 2) \div 2^2$

9. $(81 - 3) - (10 \div 2)^2$

10. $(12 - 3) \div (3 + 3) + (2 \times 2)$

11. $(8^3 + 5) \div (5 - 2)$

12. $4 + (142 - 2) + 80$

13. $1{,}084\ (43 - 40) \times 9$

14. $(85 \div 5) + 2^2$

15. $(3 \times 2) \div 2^2$

16. $200 \div 2^3$

17. $50 - (6 \times 4) + 3$

18. $800 \div (5 \times 4)^2$

19. $(320 + 40) \div 3^3$

20. $(35 + 5)^2 \div 10^2$

21. $10^3 \div 2^3 + (98 \div 2)$

22. $15 + 4 - 8 + 7$

23. $(7 \times 7) - (6 \times 3)$

24. $972 \div 2^3$

25. $2 \times 10^3 + (6 \div 3)$

26. $5^2(2^3) \div 4^2$

27. $(45 - 15) \div (2 + 1)$

28. $189 - 9^2 + 234$

Compare >, <, or =

29. $12 + (28 - 5)$ ____ $(12 + 28) - 5$

30. $39 - (16 \div 2)$ ____ $(36 \div 3) + 3^3$

31. $(81 \div 3^3) \times 6$ ____ $(22 \times 4) - 22$

32. $2{,}480 - (133 + 12)$ ____ $(2480 - 133) + 12$

33. $(9 + 3) - (6 + 2)$ ____ $2 + 3$

34. $(4 \times 4) \times 9 \times (3 \times 3)$ ____ 81

35. $(5 + 20) \times 4$ ____ 10^2

Solve if x = 3 and y = 5

36. $x^3 + (y - x)$

37. $4(x + y) - y^2$

38. $(x^2-2) \div x$

39. $(x - y)^3$

40. $2x + y^2$

41. $3x + 5y$

42. $10y - 5x$

43. $(x + y)^3 - 25$

Unit 3.1 5.NR.2.1 Multiply and divide multi-digit whole numbers to solve relevant mathematical problems. (DOK 2)

Imagine going to a carnival with your best friend in town. Your parents agree that you could spend two hours at the carnival, but they only gave you each a $20.00 bill to spend. In order to make it interesting, your parents agreed to give an additional $20.00 bill to the one that spent their money the wisest, yet had the most fun. Just like you guessed, there was a $2.00 gate fee. Helping veterans seemed like a good thing to do, so both happily agreed to pay and enter the carnival.

Immediately, Josh taps you on the shoulder and points to the hot dog stand!

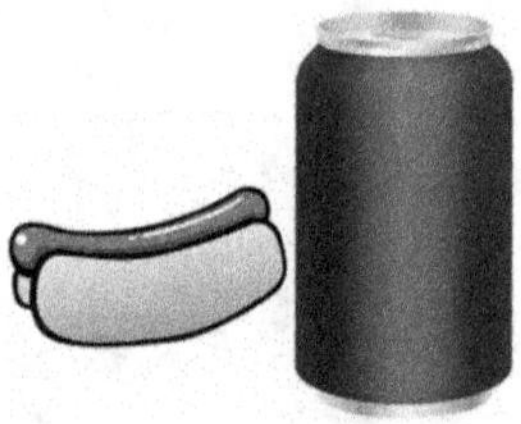

Of course, his stomach is always growling! The cashier said that the hot dog was $1, the soda was $1.50, and a bag of popcorn $1. Now, you decide to create the rest of your journey. Rides are $2, and games $1. Take ten minutes to work with a desk partner and tell your story. PS, your teacher has agreed that the best story has a free homework pass! Good luck!

For a softball celebration of increased improvements over the year, the athletic department has asked Pollo Bueno to cater.

Coach Croft has chosen the package deal. Each table is $150 per table. If there are 200 people attending, and each table seats 8 people. How much will Pollo Bueno charge the school?

Coach Croft has calculated that the total for the banquet will be (120 + 250) x 8 = $

Coach Stanley has calculated that the total for the banquet will be (250 / 8) x 120 = $

Coach Gaines has calculated that the total for the banquet will be (250 x 8) + 120 = $

Who is correct? And why?

Chapter Quiz: Properties of Numbers

Understanding the Property of the Number Zero

1. Which property is demonstrated by the equation 425 = 210 + 215?

A) Identity Property

B) Commutative Property

C) Distributive Property

D) Associative Property

2. What is the name of the property where any number multiplied by zero equals zero?

A) Zero Property of Multiplication

B) Identity Property

C) Distributive Property

D) Commutative Property

3. The equation 2×5 = 5×2 is an example of which property?

A) Zero Property of Multiplication

B) Commutative Property

C) Associative Property

D) Distributive Property

4. Which property is illustrated by the expression (3(3×3) = (3×3)3)?

A) Commutative Property

B) Associative Property

C) Identity Property

D) Distributive Property

Use knowledge of properties to simplify the expressions.

5. Simplify 9(7) +9(10).

A) 90

B) 7

C) 84

D) 97

6. Simplify 8 times 11.

A) 80

B) 88

C) 81

D) 18

7. Simplify 25 times 2 times 4.

A) 200

B) 150

C) 100

D) 50

8. Simplify 877 times 3.

A) 2,631

B) 1,754

C) 877

D) 0

Solve the Problems

9. What is the result of 919 + 0?

A) 0

B) 919

C) 1

D) 920

10. Multiply 280 times 5.

A) 1,500

B) 1,400

C) 1,000

D) 1,200

11. Simplify 345 times 4.

A) 1,380

B) 1,200

C) 1,400

D) 1,500

12. Add 3,017 + 53.

A) 3,070

B) 3,000

C) 3,100

D) 3,060

13. Subtract 1,860 - 75.

A) 1,800

B) 1,785

C) 1,750

D) 1,835

Solve the Problem Using Rounding First

14. Round and solve 29 times 810.

A) 23,000

B) 24,000

C) 25,000

D) 26,000

15. Round and solve 57 times 792.

A) 45,000

B) 47,000

C) 50,000

D) 48,000

Unit 3.2 5.NR.2.2 Fluently divide multi-digit whole numbers (up to 4-digit dividends and 2-digit divisors no greater than 25) to solve real-life problems. (DOK 2)

There are four groups of animals at the Chattanooga Zoo: birds, mammals, reptiles, and fish. If the ratio is evenly divided, how many animals are mammals if there are 120 animals at the zoo?

Answer: 120 divided by 4 equals 30.

Ex. 1 There are approximately 30 fish in the zoo. If half the fish are labeled as sharks, how many of the fish are sharks?

Therefore, 30 divided by 2 is 15.

Answer:

```
  15
2)30
- 2
  10
- 10
```

Ex. 2 There are 580 fifth graders at Jasper Elementary. The students have worked hard to earn a Math field day. There are lots of games to choose from. In order to be manageable, teachers and students agreed that the classes of competitors would be divided into groups of twenty. How many classes are in the fifth grade at Jasper Elementary?

```
     29
20)580
  - 40
    180
  - 180
```

Divide.

1. 52 ÷ 2

2. 49 ÷ 7

3. 75 ÷ 3

4. 144 ÷ 4

5. 175 ÷ 5

6. 224 ÷ 4

7. 121 ÷ 11

8. 144 ÷ 12

9. 2,400 ÷ 8

10. 3,632 ÷ 8

11. 4,545 ÷ 15

12. 3,636 ÷ 12

Divisibility

Let's consider what divisibility is. Well, divisibility is when a number can be divided by another number evenly. Divisibility rules are shortcuts for determining if a number is divisible by another number without using long division. For example, if a number ends with a 0 or 5. It is divisible by 5.

Common Divisibility Rules

A number is divisible by ...	If	Divisible	Counter Example
2	The last digit is a 0 or divisible by 2	104	101
3	The sum of the digits is divisible by 3	150	110
4	The last two digits are divisible by 4	288	139
5	The last digit is a 0 or 5	120	121
6	The number is divisible by 2 and 3	36	37
8	The number made of the last three digits is divisible by 8	1440	1441
9	The sum of the digits is divisible by 9	81	82
10	The last digit is 0	360	361

Determine if these numbers can be divided by

2, 3, 4, 5, 6, 9, or 10.

1. 920

2. 525

3. 837

4. 450

5. 763

6. 900

7. 322

8. 2,789

9. Will 2 go into these numbers? 920, 765, 360, 872, 954, 688

10. Will 5 go into these numbers? 720, 385, 981, 453, 520, 930

11. Will 4 go into these numbers? 850, 120, 840

12. Will 3 go into these numbers? 90, 330, 165

13. Will 10 go into these numbers? 350, 728, 830

14. Will 6 go into these numbers? 269, 860, 345

15. Which number between 1 and 50 can be divisible by 2,3,4,6 and 9?

16. Which number between 1 and 70 can be divisible by 2,3,4,5,6 and 10?

Challenge: Let's look at larger numbers using the same divisibility rules.

Are the following numbers divisible by 2,3,4,5,6 and 10?

List which numbers they are divisible by.

17. 1,440

18. 25,689

19. 2,384,232

20. 9,999

21. 5,420,830

22. 649,921

23. 842,235

24. 34,930

25. 75,555

Unit 3.3 5.NR.5.1 Write, interpret, and evaluate simple numerical expressions involving whole numbers with or without grouping symbols to represent real-life situations. (DOK 2)

Ex. 1 Write a simple numerical expression. Five times the sum of 45 and 15.

5(45 + 15)

Ex. 2 Three times the difference in 5 puppies and 2 kittens.

3 (5p - 2k)

Ex. 3 Explain whether the product of 25 and 30 is equal to the Area Model of 35 and 20.

Ex. 4 Simplify. 4x + 3y , when x = 8 and y = 5.

Write the following numerical expressions with whole numbers.

1. The sum of ten and four.

2. The difference between seventeen and twelve.

3. Four times three

4. Twenty divided by five.

5. The sum of seven and three multiplied by two.

Interpret the numerical expression.

1. The difference of twice as many cats as dogs.

2. The total of 5 quartz and 5 rubies added to the sum of 3 quartz and 2 rubies.

3. The sum of four chocolate candies and six peppermints divided among two students.

4. The product of $10 per hour and the eight hours that Scott worked on Friday.

5. Double the perimeter of a sandbox with a length of ten feet and a width of five feet.

6. Explain whether the product of 20 and 30 is equal to the area model of 4 and 5 and 6 and 10.

7. Explain whether the product of 10 and 6 is equal to the area model of 3 and 2 and 2 and 5.

Evaluate the following expressions, given c = 6, d = 4, and f = 5:

8. $15d - 2f$

9. $8 (f + d)$

10. $7(c + d) / f$

11. $3 (c + d + f)$

12. $(c \cdot d) / 2$

13. $(d \cdot f)(d \cdot f)$

14. $(d + f)(d + f)$

15. $(c + d + f) / 6$

Unit 3.4 5.MDR.7.2 (DOK 2)

There are some real-world applications to multiplying and dividing whole numbers. Let's consider a few examples.

Example 1: Kaleb and Josh are working the gate at a local football game. Tickets are $10 for adults and $8 for students. Josh's mom and brother are purchasing tickets to watch the game. They want to buy ten additional tickets for some neighbors. There are 6 adults and 4 students. What is the total that Josh's mom paid for tickets to the game? 10 + 8 + 6(10) + 4(8) is an expression to find the amount of money needed for tickets. 18+60+32 can be simplified to 50+60 = $110.

Example 2: A group of bikers want to ride on Dahlonega Trails. If they paid a total of $160 for tickets to the resort, then how much did each biker pay if there are a total of 8 bikers? Well, $160 ÷ 8 = $20 a person.

Simplify the following.

1. There are three horses at the North Georgia Nature Center. If each horse eats about twenty pounds of hay a day, how many pounds of hay do they eat all together in a week?

2. If a small bale of hay costs $10, how many bales of hay would it take to feed all three horses in a week?

3. If a small bale of hay costs $10, how many bales of hay would it take to feed all three horses in a month?

4. A farmer wants to be generous and gives the Nature Center 3 large round bales of hay that sell for $100. How much money did he save the center?

5. The farmer owns a poultry farm that sells eggs. Since his family is retired, he does it to stay busy. He offers his neighbors a dozen eggs for $2. If he has 300 layers that produce 250 good eggs per day, how many eggs can he gather in a month?

6. If it costs $1 to feed four hens a week, how much profit is the farmer making assuming that the hens are healthy and continue to lay around 250 good eggs per day?

7. During the summer months, the farmer noticed that his hens laid fewer than 100 eggs. This puzzled Mr. Jones, so he asked his local county extension agent for ideas as to why the decrease in egg production. What are two reasons that the extension agent may say to help Mr. Jones understand chickens?

8. Carlos wants to save money for a new dirt bike. He makes a chart of the money he receives from his two after-school jobs. He cuts grass and charges his neighbors $40 to cut their lawns. He washes cars and charges his neighbors $20 to detail their cars. If he spends $20 for gas each week and $10 for cleaning supplies each week, how long will it take him to save money for a $1,000 dirt bike?

Unit 3 Review (DOK 2)

Shade the answers to the solved problems on the number grid.

1	2	3	4	5	6	7	8	9	10
11	12	13	14	15	16	17	18	19	20
21	22	23	24	25	26	27	28	29	30
31	32	33	34	35	36	37	38	39	40
41	42	43	44	45	46	47	48	49	50
51	52	53	54	55	56	57	58	59	60
61	62	63	64	65	66	67	68	69	70
71	72	73	74	75	76	77	78	79	80
81	82	83	84	85	86	87	88	89	90
91	92	93	94	95	96	97	98	99	100

Simplify.

1. $9 \cdot 9$ **2.** $2(20+21)$ **3.** $3(20+5) - 2$

4. $3(7 \cdot 3)$ **5.** $(2 \cdot 25) + 4$ **6.** $(3^2 \cdot 5)$

7. $(3 \cdot 3)(2 \cdot 2)$ **8.** $(3)(3)(3)(1)$ **9.** $(3+6)(2)$

10. $4+4+3+2$ **11.** $(2)(2)(2)(3)$. **12.** $(9 \cdot 5) - (2 \cdot 5)$

13. 4(12) - 2

14. 5(7+4)+2

15. (2+12)(2+2)

16. (8 · 10) - 1

17. (3+12) (2·3)

18. 1+1+1+2

19. (2)(2)(2)+7

20. (2+3)(9-4)

21. 100 - 6.

22. 4(7·3)

23. (5·5)(4-1).

24. 80 - 4

25. (3·3·2·5)-3

26. 10·10-3

Evaluate the expressions when a = 2 , b = 3, and c = 4

27. 3c - ab

28. bc / a

29. John makes abc dollars per hour. How much will he make in a 40 hr week?

30. John wants to save money for a computer that costs $500; how many 40-hour weeks will he need to work to make more than $500?

Unit 3 Test Review:

5. NR.2.1 Fluently multiply multi-digit (up to 3-digit by 2-digit) whole numbers to solve real-life problems. 5.NR.5.1 Write, interpret, and evaluate simple numerical expressions involving whole numbers with or without grouping symbols to represent real-life situations. 5.NR.2.2 Fluently divide multi-digit whole numbers (up to 4-digit dividends and 2-digit divisors no greater than 25) to solve practical problems. 5.MDR.7.2 Ask questions and answer them based on gathered information, observations, and appropriate graphical displays to solve problems relevant to everyday life.

Properties:

Associative: (ab)c = a(bc) **Ex.** (3*5)*4 = 3(5*4)

Commutative: a * b = b * a **Ex.** 5*6 = 6*5

Distributive: a(b+c) = ab + ac **Ex.** 2(4+8) = 2(4)+2(8)

Identity: a*1 = a **Ex.** 3*1 = 3

Zero Property: a*0 = 0 **Ex.** 9*0 = 0

Determine the Property:

1. 4*25 = 2*10+8*10

2. A number multiplied by zero.

3. 2(5) = 5(2)

4. 3*(3*3) = (3*3)*3

Explain the property used.

5. 9*0*7 = 0

6. 2(81) = 2(9*9)

7. 25*2*4 = 4*2*25

8. 3(5+5) = 3(5)+3(5)

9. 7*1 = 7

10. 8(7)3 = 7(8)3

Multiply.

11. 91×9

12. 280×5

13. 345×4

14. 3,017×33

15. 300×50

16. 245×81

17. 305×86

18. 1,860×75

Divide.

19. 275/5

20. 1,800/25

21. 1,492/4

22. 8,459/3

23. 1,565/15

24. 2,764/2

25. 3,625/25

26. 5,732/2

Multiply using Part-Whole.

27. Marc is wanting to find the area of his field. It is 45 yards in length and 17 yards in width. He chooses Part-whole thinking to help him multiply. Is his strategy right?

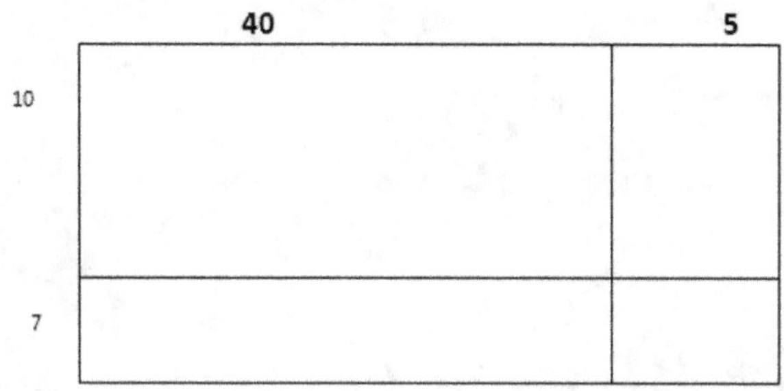

28. Create a part-whole for multiplying 25 by 15.

Divide using Partial Quotient:

29. How many times can 5 go into 145?

We know that 5 x 20 = ?

We know that 5 x 9= ?

Therefore, 145 divided by 5 = ?

30. Luis wants to know how many groups of 5 members can fit into the plane that holds 235.

He thinks: 5 x 20 = 100

5 x 20 = 100

7 x 5 = 35

Therefore, there should be 47 groups of 5 able to fit on the plane.

Unit 4:

Building Fraction Understanding

Unit 4 Vocabulary

These are the vocabulary words to be discussed in this unit. Try to have fun with them and discuss them with your learning partners and stakeholders, i.e., parents, students, coaches.

Addition/Add: The process of combining two or more numbers or terms.

Benchmark Fraction: A fraction used as a reference point, in order to compare something.

Common Denominator: A number that is shared by other fractions.

Denominator: The bottom number of a fraction.

Difference: What remains when subtracting two numbers or terms?

Estimate: The process of finding a number close to the answer.

Equivalent: Another name for equal or equal amount.

Fraction: A way to describe part of a whole.

Fraction greater than 1: A fraction whose numerator is larger than the denominator.

Fraction less than 1: A fraction whose numerator is less than the denominator.

Mixed number: A fraction that is combined with a whole number.

Numerator: The top part of a fraction.

Reasonableness: The process of asking if a number makes sense.

Subtraction/Subtract: The process of taking a number away from another number.

Unit Fraction: A fraction with the numerator being 1.

Unlike Denominator: When the denominators of two fractions are not the same.

Create a story with the vocabulary. Draw pictures, if time permits. Share with friends.

Unit 4.1 5.NR.3.2 Compare and order up to three fractions with different numerators and/or different denominators by flexibly using a variety of tools and strategies. (DOK 3)

Jeff and Jordan are on a mission. After successfully navigating through a time continuum, they meet three hungry turtles. There are trees all around, but the turtles seem to be interested in their shiny sneakers. Through careful estimation, it seems that the year is 1,000. There are clear blue rivers, and smoke can be seen rising above the trees in the far distance.

Jordan thinks that the smoke is from a natural source, like a lightning strike, but Jeff isn't buying it. He says, "Let's go half the distance today, 1/4 tomorrow, and 1/3 of the way since there seems to be some mountains in the way." Of course, they left their phones in the time machine; after all, they were amazed to believe it worked at all. Jeff says, "Let's walk the smallest part today to get a feel of the land," remembering to mark where they landed. The middle part tomorrow, and the largest part the next day, since they would be familiar with the land at that point.

Example 1: So, given the ratios: 1/2, 1/4, and 1/3, let's order them from least to greatest by comparing fractions. First, we find the LCM in order to get a common denominator. Let's look, there is a 2, a 4, and a 3. We know that each of these numbers divide evenly into 12. Therefore, the LCM is 12.

1/2 = 6/12

1/4 = 3/12

1/3 = 4/12

Example 2: Now we can order them: 1/4, 1/3 and 1/2. Walking on, they meet an Indian. He has determined that they are lost and chooses to help them. Strong Wolf offers the two travelers 1/2 a melon, 3/4 a basket of corn, 1/10 a palm of oil, and 3/5 a basket of fish. Order the food based on their ratios.

Now, we have to use the LCM to create equivalent fractions. Once they have the denominator, we can order the fractions.

1/2 = /60

3/4 = /60

1/10 = /60

3/5 = /60

Now we can order 1/10, 1/2. 3/5. and 3/4.

Wouldn't it be neat to see what the Ocmulgee Indian Mounds were built from? How were they created, and what purposes did they serve before the European settlers came to these shores? Study your history and discover that one mound was partially destroyed by the railroad projects. If Jeff and Jordan set their time machine ahead to 2,025, they would be surprised to know that the Muscogee have made an effort to purchase the land around the mounds and are in the process of rebuilding the monuments to make them into National Monuments.

Arrange the following fractions from least to greatest.

1. 3/4, 7/12, 4/6

2. 4/3, 3/4, 5/1

3. 5/12, 7/20, 11/24

4. 3/15, 7/5, 1/3, 1/15

5. 5/9, 3/5, 12/45, 8/9

Arrange the following fractions from greatest to least.

6. 4/9, 6/12, 9/6

7. 1/4, 1/5, 1/9

8. 2/3, 5/6, 4/9

9. 6/7, 2/14, 4/7. 5/14

Unit 4.2 Scavenger Hunt, Fractions (DOK 2)

New mission: you are given a set of numbers 0 through 9, and you are tasked with explaining to your partner that you ate 3/8s of a pizza last night. You pick up a 3 and an 8 from the pile of numbers. You explain to your partner that you have eaten a fraction of the whole pizza. She gives you a stare, but nods in agreement once you use your pencil to make a fraction bar to separate the numerator and denominator. You see some marbles lying on the desk beside you, and you make a pictorial fraction showing how much pizza you have eaten and how much is left. Your teacher hears the discussion and suggests that you draw a diagram showing the fraction. She explained that drawings are a good way to suggest knowledge and information.

Your partner then explains that she is beginning to understand since her brother had eaten half of her 10-piece box of chicken nuggets. Should be happy that he didn't eat them all.

Given 5 minutes to talk among yourselves, find 5 examples of fractions or ratios in the classroom. For instance, 7/20 desks are sitting closer to the front of the room than the rest of the desks. Another example would be that there are 5 out of 20 Math books on the bookshelf. Or 3 of the 10 computers are a different color.

Description of Fraction	Drawing of the Fraction	Numerical Fraction

With recognizing fractions, there may come a time where you will need to order these fractions from least to greatest or greatest to least.

With a partner and a set of numbers create three fractions with the ten numbers given to you.

Ex. 1 Given 3/10, 2/10, and 6/10, order the fractions from least to greatest. Since the denominators are the same, we just have to look at the numerators. We know that 2 < 3 and 3 < 10, so we can determine that the order 2/10, 3/10, and 6/10 is correct.

Ex. 2: Given 1/2, 3/5, and 2/3, order the fractions from least to greatest. (If they had common denominators, we would simply look at their numerators to determine the order.) 1/2, 3/5. and 2/3. Yet, we see that the denominators need to be changed since they are different. Using the LCM, we will look at their denominators in order to line them up from least to greatest.

$$\frac{1}{2}=\frac{10}{30} \qquad \frac{3}{5}=\frac{18}{30} \qquad \frac{2}{3}=\frac{20}{30}$$

We find the common denominators. Now, we can order the fractions from least to greatest. We have determined that 1/2 < 2/3 < 3/5.

Exercises: Order the following fractions from least to greatest.

1. 3/5, 2/5, 1/5, 4/5

2. 1/8, 6/8, 4/8, 2/8

3. 2/12, 4/6, 2/3, 1/4

4. 3/18, 2/9, 5/3, 4/2

5. 7/12, 2/3, 9/8. 3/4

Order the following fractions from greatest to least.

6. 9/10, 6/5, 7/10, 3/5

7. 6/9, 5/18, 3/9, 7/18

8. 1/12, 9/4, 5/24, 8/24

9. 5/3, 6/9, 1/3, 2/9

Estimating with Fractions

In adding fractions, we can use rounding to estimate with fractions.

Example 1: Estimate the sum:

$\frac{5}{6} + \frac{8}{12}$

We can round fractions to

0, 1/2, or 1 in order to estimate.

<----|------|------|------|------|------|------|------|------|------|------|------|------|--->

$0 \quad \frac{1}{12} \quad \frac{1}{6} \quad \frac{3}{12} \quad \frac{1}{3} \quad \frac{5}{12} \quad \frac{1}{2} \quad \frac{7}{12} \quad \frac{2}{3} \quad \frac{9}{12} \quad \frac{5}{6} \quad \frac{11}{12} \quad 1$

Determine whether the fraction is closest to 0, 1/2, or 1. By looking at the number line, we can see that 5/6 is closer to 1. We can tell that 8/12 is closer to 1/2. So by understanding this, we can estimate the sum of 5/6 and 8/12 is close to 1 1/2.

Estimate the sum or difference.

1. $\frac{1}{4} + \frac{8}{10}$

2. $\frac{7}{8} - \frac{2}{5}$

3. $\frac{4}{5} - \frac{1}{3}$

4. $7\frac{1}{3} - 4\frac{4}{5}$

5. $8\frac{5}{12} + 4\frac{1}{2}$

6.NR.3.3 Model and solve problems involving addition and subtraction of fractions and mixed numbers with unlike denominators.

1. $\begin{array}{r} 3\frac{1}{2} \\ +\ 2\frac{1}{4} \\ \hline \end{array}$

2. $\begin{array}{r} 5\frac{3}{8} \\ +\ 5\frac{3}{4} \\ \hline \end{array}$

3. $\begin{array}{r} 8\frac{2}{3} \\ +\ 1\frac{1}{6} \\ \hline \end{array}$

4. $\begin{array}{r} 9\frac{1}{10} \\ +\ 12\frac{3}{20} \\ \hline \end{array}$

5. $\begin{array}{r} 2\frac{4}{9} \\ +\ 15\frac{1}{3} \\ \hline \end{array}$

6. $\begin{array}{r} 8\frac{3}{4} \\ +\ 7\frac{5}{6} \\ \hline \end{array}$

7. $\begin{array}{r} 13\frac{1}{2} \\ +\ 12\frac{3}{7} \\ \hline \end{array}$

8. $\begin{array}{r} 14\frac{2}{5} \\ +\ 6\frac{3}{10} \\ \hline \end{array}$

Unit 4 Subtracting fractions (with borrowing)

(DOK 3)

When subtracting Mixed fractions, sometimes you have to borrow in order to subtract since the numerator being subtracted is greater than the denominator being subtracted from.

Example 1.

$$\begin{array}{r} 10\frac{3}{10} \\ -7\frac{6}{5} \\ \hline \end{array}$$

Get common denominator.

$$\begin{array}{r} 10\frac{3}{10}=10\frac{3}{10} \\ -7\frac{6}{5}=7\frac{12}{10} \\ \hline \end{array}$$

Since 12 > 3, we need to borrow by taking 1 away from the whole number and adding to the fraction. 10 - 1 = 9. and 10/10 added to 3/10 is 13/10.

$$10\frac{3}{10}=9\frac{13}{10}$$

Now, we can subtract.

$$\begin{array}{r} 9\frac{13}{10} \\ -7\frac{12}{10} \\ \hline 2\frac{1}{10} \end{array}$$

Therefore, when we subtract the mixed fraction, we get .

Subtract the following fractions:

1. $\begin{array}{r} 16\frac{2}{10} \\ -3\frac{1}{5} \\ \hline \end{array}$

2. $\begin{array}{r} 8\frac{5}{6} \\ -3\frac{3}{8} \\ \hline \end{array}$

3. $\begin{array}{r} 20\frac{4}{30} \\ -15\frac{5}{6} \\ \hline \end{array}$

4. $\begin{array}{r} 18\frac{8}{15} \\ -11\frac{9}{10} \\ \hline \end{array}$

Spiral Quiz

Compare the following numbers. > , < , =

1. 56,804 ________ 54,980

2. 34,765.35 ________ 33,899.65

3. 18,136 ________ 18,097

What digit is in the following spaces?

4. 75,824 Which digit is in the thousands place?

5. 78,524.23 Which digit is in the hundredths place?

6. 79,532.86 Which digit is in the ten thousands place?

7. 82,739 The number 2 is in the ___________ place.

8. 738,201 The number 2 is in the ___________ place.

Add.

9. $\frac{3}{4} + \frac{7}{20}$

Subtract.

10. $\frac{5}{8} - \frac{1}{2}$

Unit 4.3 5.MDR.7.2 Ask Questions, Animal Adventures (DOK 2)

Welcome to the neighborhood of Fractionville, where animal friends embark on exciting mathematical fun!

Bonnie the Boxer Bulldog loves parties, and today, she's throwing a feast for her friends. She's preparing delicious peanut butter cakes for everyone. Each cake is divided into equal parts.

1. Bonnie has baked 2 peanut butter cakes. Each cake is divided into 8 slices. If Milo eats 5/8 of a cake, how many slices did he eat?

A) 2 slices

B) 3 slices

C) 4 slices

D) 5 slices

2. After the party, there are 3/8 of a cake left. Bonnie wants to share the remaining cake equally between herself and Chopper. How much cake will each get?

A) 1/8

B) 2/8

C) 3/8

D) 5/8

Fifi the Cat and Max the Dog love playing in the yard. Today, they decide to compare how much of the yard they've each explored.

3. Max has explored 3/4 of the yard, while Fifi has explored 2/3 of it. Who has explored more of the yard?

A) Max

B) Fifi

C) The same amount

D) Cannot be determined

4. To complete the yard, Max needs to explore another 1/4 of it. How much of the yard will he have explored in total?

A) 1/2

B) 1

C) 1 1/4

D) 1/8

The animals gather at the watering hole, where Fifi and Max decide to collect water for their garden.

5. Fifi collects 1/2 of a bucket of water, and Lila collects 5/8. How much water do they have together?

A) 7/8

B) 5/8

C) 1 1/8

D) 1/4

6. After watering the garden, they realize they've used 1/8 of their water. How much water do they have left?

A) 1/2

B) 3/8

C) 5/8

D) 7/8

7. Mutts has 3/6 of a watermelon, and Bella has 1/6. If they combine their watermelon, how much do they have?

A) 7/6

B) 2/3

C) 1 1/6

D) 1/2

8. Fifi brought 3/5 of a carrot and gives 1/5 to Mutts. How much carrot does Fifi have left?

A) 1/5

B) 2/5

C) 4/5

D) 5/5

9. The yard has 8 toy balls. Chopper and Max each play with 5/8 of the balls. How many of the toy balls do they play with together?

A) 3/8

B) 6/8

C) 1

D) 9/8

10. After a while, Fifi joins the game and plays with 1/8 more of the toy balls. How much of the toy balls are being played with now?

A) 1

B) 7/8

C) 5/8

D) 8/8

11. Bonnie finds out she has 3/12 of a chew toy, and Max gives her another 7/12. How much of the chew toy does Bonnie have now?

A) 9/12

B) 1

C) 3/4

D) 5/6

12. Chopper needs to walk 3/5 of a mile to reach his home. He has already walked 1/5 of a mile. How much further does he need to walk?

A) 1/5

B) 2/5

C) 3/5

D) 4/5

Quiz: Fractions

1. Which fraction is larger?

a) 3/4

b) 5/12

c) 1/2

d) 9/12

2. Order the following fractions from smallest to largest: 1/3, 2/5, 3/6.

a) 1/3 ,3/6, 2/5

b) 2/5, 3/6, 1/3

c) 3/6, 2/5, 1/3

d) 2/5, 1/3, 3/6

3. Add 1/4 + 2/3

a) 3/7

b) 11/12

c) 3/5

d) 7/12

4. Subtract 9/12 - 4/6

a) 5/12

b) 4/6

c) 2/3

d) 1/12

5. Which of the following is equal to 11/12?

a) 5/12 + 1/2

b) 3/12 + 3/4

c) 1/3 + 1/4

d) 5/8 + 3/4

Problem Solving with Graphical Displays

6. A bar graph shows the favorite fruits of a class. If 20 students like apples, 24 like bananas, and 16 like oranges, which fruit is the most popular?

a) Apples

b) Bananas

c) Oranges

d) Cannot be determined

7. A circle graph shows that 1/4 of the students walk to school, 1/3 ride a bike, and the rest take the bus. What fraction of students take the bus?

a) 7/12

b) 5/12

c) 9/12

d) 1/5

8. If there are 200 students, how many students walk to school?

a) 90

b) 85

c) 50

d) Cannot be determined

9. A survey of the 200 students said that 1/2 like chocolate, 1/4 like vanilla, and 1/10 like strawberry. Which student group likes the most flavors?

a) Chocolate

b) Vanilla

c) Strawberry

d) Cannot be determined

10. Which of the following fractions are equal to 1/4?

a. 2/10
b. 1/5
c. 9/12
d. 3/12

11. Students reported that they read books over the summer. The list is as follows:

Coach Craig's room: 3.5, 4.25, 7, 1.4, 2.6, 3, 4, 1

Coach Fulton's room: 1 1/2 , 2 1/4, 3 1/2, 2, 4, 3 1/2, 1

What is the average number of books read in each class?

12. Sketch on a number line.

13. If both classes were combined, what is the average number of books read over the summer?

Unit 4 Review

5. NR.3.2 Compare and order up to three fractions with different numerators and/or different denominators by flexibly using a variety of tools and strategies. 5.NR.3.3 Model and solve problems involving addition and subtraction of fractions and mixed numbers with unlike denominators. 5.MDR.7.2 Ask questions.

Compare fractions Using >,<, and =.

1. $\frac{1}{2}$ ___ $\frac{1}{4}$

4. $\frac{7}{8}$ ___ $\frac{6}{8}$

2. $\frac{3}{5}$ ___ $\frac{2}{9}$

5. $\frac{4}{15}$ ___ $\frac{5}{12}$

3. $\frac{5}{8}$ ___ $\frac{11}{16}$

6. $\frac{27}{50}$ ___ $\frac{53}{100}$

Order fractions.

Given three fractions, order them from least to greatest.

7. $\frac{3}{5}, \frac{7}{15}, \frac{2}{10}$

10. $\frac{3}{15}, \frac{7}{9}, \frac{1}{10}$

8. $\frac{2}{5}, \frac{7}{10}, \frac{2}{30}$

11. $\frac{4}{9}, \frac{2}{15}, \frac{3}{20}$

9. $\frac{1}{8}, \frac{3}{4}, \frac{5}{12}$

12. $\frac{3}{6}, \frac{1}{15}, \frac{7}{10}$

Add or Subtraction.

13. $\frac{3}{5}+\frac{2}{10}$

14. $\frac{13}{25}+\frac{7}{15}$

15. $1\frac{5}{8}+3\frac{12}{20}$

16. $\frac{43}{50}-\frac{7}{15}$

17. $8\frac{23}{25}-4\frac{2}{20}$

18. $\frac{96}{100}-\frac{9}{50}$

Add or subtract.

19. Luis wants to determine the distance around the field. He knows that the width measured 150 7/8 ft and the length measured 280 3/4 ft. What is the perimeter of the field?

20. Marc wants to find out how much pizza is left. He had ordered 18 pizzas, and his friends ate 9 5/8 pizzas. How much pizza is left over?

Unit 5:

Making Sense of Fraction Multiplication and Division

Unit 5: Vocabulary

These are the vocabulary words to be discussed in this unit. Try to have fun with them and discuss them with your learning partners and stakeholders, i.e., parents, students, coaches.

Denominator: The bottom part of a fraction.

Division: Decomposing a quantity into equal-sized groups.

Fraction: A way to describe part of a whole.

Equation: A mathematical statement showing equality of two expressions.

Equivalence: Two things having the same value even though the appearance may be different.

Expression: A mathematical statement containing terms and operations.

Mixed Number: A fraction that is combined with a whole number and a fraction.

Numerator: The top part of a fraction.

Partition division (or fair-sharing): A method of breaking larger numbers into smaller numbers.

Partial Product: A method of multiplying a part of a number by a part of another number.

Partial Quotient: A method of dividing a part of a number by a part of another number.

Product: Another word for multiplication.

Properties of operation: Rules that govern how math operations behave.

Quotient: Result from dividing two numbers.

Remainder: An amount, or number, leftover from dividing another number.

Unit Fraction: A fraction with the numerator being 1.

Create a crossword with the vocabulary. Use the definitions for the questions, clues, of the crossword.

Unit 5.1 5.NR.3.1 Explain the meaning of a fraction (DOK 2)

In order to explain the meaning of a fraction as division of the numerator by the denominator, let's consider something we know something about, say money. We know that there are 4 quarters in a dollar. A quarter is of a dollar, and we can say that $1 divided by 0.25 is 4. If we consider a year, we know that there are 12 months in a year. If we wanted to find the number of months in a season, we would divide 12/4, since there are four seasons. We would know that there are 3 months in a season. It doesn't mean that sometimes the seasons seem longer; that would be called science.

Example 1: You've been told by your older brother that you can help work on your sister's 2023 LT1 Convertible Camaro. She wants ground effects put on. Your brother said you have to know your tools to be able to tell what a 3/8 in bolt and a 5/8 in bolt are. "Simple," you explain, "that is three parts of an inch divided into 8ths." Your cousin says, "I thought bolts were measured by 12ths or 16^{ths}."

Answer: In general, bolts are measured by width and length. Yes, bolts are measured by 8ths and 16ths width, not 12ths.

Example 2: You are helping to distribute school supplies to a neighborhood in need. You are in charge of placing the correct amount of crayon boxes into the school totes to be given to the students and families. If there are 120 boxes of crayons, how many boxes can be given to 50 students? Will there be any left over?

Answer: We know that there are 120/50 is the numerator divided by the denominator to get get 2 boxes for each student with a remainder of 20 left over. The mixed fraction answer would be .

Divide the following numerators and denominators to create fractions:

1. 12÷5

2. 13÷3

3. 15÷3

4. 8÷4

5. 7÷2

6. 24÷5

7. 36÷8

8. 400÷50

9. 100÷20

10. 121÷10

5. NR.3.4 Model and solve problems involving multiplication of a fraction and a whole number.
(DOK 2)

Multiplication: Fraction by a Whole Number

When multiplying a fraction by a whole number, the whole number can be written over 1 to change the whole number to a fraction. Remember, 4/1 is still 4.

Example 1:

$\frac{3}{4} \times 8$ Multiply:

$\frac{3}{4} \times \frac{8}{1}$ Step 1: Change whole number to a fraction. ×

$\frac{3}{4} \times \frac{8}{1} = \frac{24}{4}$ Step 2: Multiply the fractions. =

$\frac{24}{4} = 6$ Step 3: Reduce if possible. = 6

Example 2: Half of a dozen.

(Interpret expression, 1/2 times 12)

$\frac{1}{2} \times \frac{12}{1}$ Step 1: Change to fractions.

$\frac{3}{4} \times \frac{12}{1} = \frac{36}{4}$ Step 2: Multiply the fractions.

Step 3: Reduce if possible.

Multiply.

$\frac{36}{4} = 9$

1. $\frac{7}{8} \times 4$ **2.** $\frac{3}{5} \times 5$ **3.** $\frac{8}{14} \times 7$ **4.** $\frac{4}{5} \times 2$ **5.** $\frac{6}{32} \times 8$

6. $\frac{1}{2} \times 10$ **7.** $\frac{3}{5} \times 30$ **8.** $\frac{8}{14} \times 70$ **9.** $\frac{4}{5} \times 50$ **10.** $\frac{6}{32} \times 16$

11. $\frac{3}{9}$ of 27 **12.** $\frac{4}{5}$ of 80 **13.** $\frac{8}{14}$ of 7 **14.** $\frac{4}{32}$ of 8 **15.** $\frac{6}{27}$ of 9

16. Three-fourths of thirty

17. Two-fifths of sixty

18. Seven-tenths of a hundred

19. Half of a foot

20. A fourth of a foot

Fractions and Whole Number Operations Quiz

1. What is the value of 12/3?

A) 1

B) 2

C) 3

D) 4

2. What is the value of 160/4?

A) 40

B) 35

C) 20

D) 25

3. Multiply: 1/4 by 480

A) 12

B) 10

C) 120

D) 160

4. Which statement is true about multiplying a whole number by a fraction greater than one?

A) The product is less than the whole number.

B) The product is equal to the whole number.

C) The product is greater than the whole number.

D) The product is a unit fraction.

5. Simplify 360 × 2/3.

A) 12

B) 120

C) 240

D) 510

6. Divide 455 ÷ 5

A) 19

B) 91

C) 36

D) 45

7. When 3/4 is multiplied by 84, what is the product?

A) 34

B) 39

C) 61

D) 63

8. Divide 1175 ÷ 25

A) 345

B) 340

C) 45

D) 47

9. Multiply 1860 × 3/4

A) 1340

B) 1380

C) 1350

D) 1390

10. Add 1/2 + 3/8.

A) 4/8

B) 5/7

C) 6/8

D) 7/8

Unit 5.2 5.NR.3.5 (DOK 2)

Explain why multiplying a whole number by a fraction greater than one results in a product greater than the whole number, and why multiplying a whole number by a fraction less than one results in a product less than the whole number and multiplying a whole number by a fraction equal to one results in a product equal to the whole number.

When you multiply a whole number by a fraction greater than one, you are essentially taking a portion of that whole number multiple times. For example, if you have 3 multiplied by 1.5, you are essentially taking 1.5 times the value of 3, which results in a product larger than 3. This is because you are increasing the original whole number by a fraction greater than one.

Conversely, when you multiply a whole number by a fraction less than one, you are taking a portion of that whole number that is smaller than the whole. For instance, if you multiply 3 by 0.5, you are taking half of 3, resulting in a product less than 3. This is because you are reducing the original whole number by a fraction less than one.

Ex. 1 When multiplying a whole number by a fraction greater than one. 4/3 x 12. The fraction 4/3 is greater than one. We're going to multiply it by a whole number, 12. The product is 48/3 = 16.

Ex. 2 When multiplying a whole number by a fraction less than one. 3/4 x 12. The fraction 3/4 is smaller than one. We're going to multiply it by a whole number, 12. The product is 36/4 = 9.

Multiply the whole number by a fraction less than 1.

1. $4 \cdot \frac{1}{2}$

2. $12 \cdot \frac{3}{4}$

3. $16 \cdot \frac{1}{8}$

4. $15 \cdot \frac{2}{3}$

5. $24 \cdot \frac{5}{6}$

6. $72 \cdot \frac{2}{9}$

7. What have you noticed about multiplying a whole number by a fraction less than 1?

8. $4 \cdot \frac{5}{2}$

9. $30 \cdot \frac{4}{3}$

10. $36 \cdot \frac{9}{4}$

11. $90 \cdot \frac{7}{6}$

12. $45 \cdot \frac{5}{3}$

13. $140 \cdot \frac{10}{7}$

14. What have you noticed when you multiplied a whole number by a fraction greater than 1?

Unit 5.3 5.NR.3.6 Model and solve problems involving division of a unit fraction by a whole number and a whole number by a unit fraction.

(DOK 2)

Ex. 1 A farmer has been asked if he can sell different sizes of egg cartons. He has considered selling a half dozen, a dozen, and a one and half dozen egg cartons. If he chooses to sell a dozen eggs for $3, how much does it cost for one egg? Hint: find the unit price.

Ex. 2 What would be an acceptable price for the half dozen eggs considering the unit price?

Ex. 3 What would be an acceptable price for the one-and-a-half dozen egg carton?

Find the unit fraction.

1. The apple bag at the local fair was $5. How much is the unit fraction, or the price of an apple, if there were 20 apples in the bag?

2. The 12-pack of soda costs $3. How much is the unit fraction?

3. There are five bags of street corn in the carton. Each unit of street corn was $ 1 1/2. What is the carton of street corn worth?

4. Ryan's dad asked him to look at newspaper ads and the internet to find the best price for tires. His dad needs to buy 4 tires for his Kia. Ryan found an ad with the price of one tire for $200, mounted and balanced. What is the price for all four tires?

5. You decide to shop early for next year's school supplies. You have noticed a very good sale of pencils. If the unit price for one pencil is 2 1/2 cents, how much will twenty cost you?

6. If a carton of crayons costs $ 0.25, how much does one crayon cost? There are 16 crayons in the carton.

7. After you text your parents how much $ the crayons cost, they ask you to get a box of them to donate to those in need. There are 50 crayon cartons in the box. What would be the total cost of the crayons?

8. While on a visit to the local market, you see a sign that says "cinnamon rolls $3." There are ten wonderful rolls in the carton. What is the unit price of a roll?

9. A local restaurant is having a sale on cheeseburgers. One quarter-pound cheeseburger is $ 1 1/4. If you decide to surprise your class of 20 students, how much money would you need to buy all of them a cheeseburger?

10. A local thrift store is having a $5 bag sale for gently used clothes. A neighbor is going through some difficult times, and you decide to purchase a bag for her. If there are 10 shirts in the bag, how much did one cost? Was this a good deal?

Unit 5 Review

5.NR.3.1 Explain the meaning of a fraction as division of the numerator by the denominator (a/b = a ÷ b). Solve problems involving division of whole numbers leading to answers in the form of fractions or mixed numbers.

5.NR.3.4 Model and solve problems involving multiplication of a fraction and a whole number.

5.NR.3.5 Explain why multiplying a whole number by a fraction greater than one results in a product greater than the whole number, and why multiplying a whole number by a fraction less than one results in a product less than the whole number and multiplying a whole number by a fraction equal to one results in a product equal to the whole number.

5.NR.3.6 Model and solve problems involving division of a unit fraction by a whole number and a whole number by a unit fraction.

Explain the meaning of a fraction. Solve problems involving division of whole numbers.

1. A numerator divided by a denominator a/b = a÷b. For example. 30/5 = 30÷5.

2. Joan has 24 cookies that she wants to divide between her and her five friends. How many cookies can each have if each receive the same amount?

3. Kaynan has a Taco box. He bought 36 tacos for $18. How much did each taco cost? Or what is the unit fraction?

Divide.

4. 48÷12

5. 60÷5

6. 125÷6

7. 240÷18

8. 150÷20

9. 175÷15

Multiplying a whole number by a fraction. Tell whether the whole number is bigger than the fraction, results in the number being greater than the number and whether the results are smaller if the number is multiplied by a fraction less than 1.

10. $12 \times \frac{1}{4} =$ ______

11. $24 \times \frac{2}{3} =$ ______

12. $36 \times \frac{4}{3} =$ ______

13. $48 \times \frac{7}{6} =$ ______

14. $50 \times \frac{1}{10} =$ ______

15. $100 \times \frac{8}{5} =$ ______

Conclusion: When a whole number is multiplied by a fraction less than 1, the fraction is ______. When a whole number is multiplied by a fraction greater than 1, the fraction is ______

16. There are 24 hours in a day. The unit fraction of an hour is ________.

17. Caleb works at a local store. He brought home $360 a week. What is the unit fraction for hours worked at the store?

18. Conner plays a game where the unit fraction is 1/2. If he plays 20 turns. How many units does he expect to earn?

19. Travis bought a dozen eggs for $3. What is the unit fraction for an egg?

Unit 6:

Extending Place Value and Working with Decimals to Solve Problems

Unit 6 Vocabulary

These are the vocabulary words to be discussed in this unit. Try to have fun with them and discuss them with your learning partners and stakeholders, i.e., parents, students, coaches.

Decimal: A way a writing numbers as parts of a whole based on the number 10.

Decimal fraction: A fraction with the denominator is a power of 10.

Decimal number: A number that contains a decimal, usually whole, and parts of a whole.

Decimal point: A point that separates the whole and parts of a whole.

Digit: Any single number from 0 to 9 which can be used to create other numbers.

Estimate: A process of finding an answer close to the exact answer.

Fluency: The ability to solve math quickly and correctly.

Hundredth: One part of something divided into 100 equal parts, 0.01.

Part-whole strategy: A way to think about solving math problems by breaking them down and putting them back together.

Place value: A value of a digit based on the position of the number.

Round: A way of writing a number in simpler notation.

Strategic thinking: A way to plan and use mathematical steps to solve problems.

Strategy: A method used to solve problems.

Tenth: One part of something divided into 10 parts, 0.1.

Thousandth: One part of something divided into 1,000 parts, 0.001.

Create a story with the vocabulary. Share with friends.

Unit 6.1 5. NR.4.1 Read and write decimal numbers (DOK 1)

Some fields of study require an understanding of being able to read and write decimal numbers to the thousandths place and beyond. Obviously, Math is one of them. Science can be another. Even Finance.

Ex. 1 There is an experiment in Physics that requires the use of numbers in Standard Form. Mr. Jones, has given you a paper to prepare your notes before the experiment. He said that you will need to write the following numbers in order to calculate the distances measured by launching the tennis ball into space. The first time, it was thrown by a classmate, and the measure was two thousand three hundred and fifty-six centimeters.

Ex. 2 The second time, it was launched by a catapult that your friend, Mike, had created. That distance was double that of being thrown by hand. What number would that be in standard form?

Ex. 3 There's always one. Not to be outdone, your cousin, Randy, brings a launcher from the batting cage. Its distance is twice the length of the ball launched by the catapult, plus five hundred and thirty-five hundredths cm. What number would that be in standard form?

Write the following numbers in standard form:

1. Five hundred thirty-three and seventy-two hundredths.

2. Four thousand one hundred fifty-seven and eighty-one hundredths.

3. One thousand eight hundred sixty-one and nine-tenths.

4. Twenty thousand four hundred and ninety-six hundredths.

5. Eighty-two thousand nine hundred seventy-four and three hundred twenty-five thousandths.

Write the following numbers in expanded form:·

6. Thirty thousand three hundred eighty-two and seven tenths.

7. Seventy-three thousand six hundred twenty-four and one hundred eighty-two thousandths.

8. Ten thousand four hundred thirteen and seven tenths.

9. Seventeen thousand nine hundred forty-two and sixty-three hundredths.

10. Five thousand three hundred eighty-six and forty-one hundredths.

Given the following numbers, write them in word form:

11. 4,525.73

12. 39,872.871

13. 97,536.4

14. 17,843.465

15. 34,714.38

Given the following numbers, change from expanded to standard form:

16. (8 · 1000) + (5 · 100) + (3 · 10) + (9 · 1) + (2 · 1/10) + (4 · 1/100)

17. (9 · 10,000) + (7 · 1,000) + (3 · 100) + (2 · 10) + (4 · 1/100) + (5 · 1/1000)

18. (5 · 1000) + (6 · 100) + (3 · 10) + (4 · 1/10) + (7 · 1/100) + (2 · 1/1000)

19. (3 · 10,000) + (8 · 100) + (2 · 10) + (6 · 1/10) + (7 · 1/100) + (8 · 1/1000)

20. (4 · 10,000) + (2 · 1,000) + (3 · 100) + (5 · 10) + (3 · 1) + (8 · 1/10)

Given the following numbers change from expanded to standard form:

[illegible]

Unit 6.2 5. NR.4.2 Read, write, and compare decimal numbers to the thousandths place, and round and perform operations with decimal numbers to the hundredths place to solve real-life mathematical problems. (DOK 2)

In the world today, we can use decimals to help measure the amount of gasoline pumped into a car to get to work. Decimals can help you understand the amount of money that a vehicle costs. Whether it is a 1985 Toyota Camry or a 2024 Lambo, decimals change the conversation. Don't believe me? Just try it. "Hey, Dad, can I have an 85 Camry or a new Lambo?"

Buying a new or used car can be a challenge. Unless the car is bought with cash, you may be asked several questions by the finance manager. Besides the obvious, "Did you find what you were looking for?" You may be asked if you have a job, and how much you can put down for a down payment. The more you can put down, the better because it will help you reduce your payments in the long run. You may be asked if the calculated monthly payment is something you can afford.

This is similar to buying a home too. Homes today require money and upkeep, which costs more money. A lot of people get a mortgage to pay for a home. Like financing cars, there is a percentage rate on paying monthly payments for a home.

Read the following decimals:

1. 200.50

2. 150.25

3. 600.1

4. 8.765

5. 92.2333

Write the following decimals.

6. Three hundred and 6 tenths

7. Eight thousand thirty-five and sixty-three hundredths

8. Seventy-five and eighty-two hundredths

9. Sixty-eight and four thousandths

10. Nine hundred thousand forty-six and seventy-three hundredths

Compare the following: Use > , < , and = .

11. 23.17 ___ 8.761

12. 875.2 ____ 92.62

13. 941.763 ____ 7,839.12

14. 48,456.34 ____ 834,528.4

15. 36.55 _____ 36.550

Round to the nearest whole number

16. 364.35

17. 849.5

18. 4,902.34

19. 5,748.43

20. 249.7

Round to the nearest tenth

21. 458.9

22. 384.54

23. 5292.49

24. 484.54

25. 49.76

Round to the nearest hundredth

26. 3729.534

27. 902.123

28. 76.872

29. 1203.678

30. 8530.258

Unit 6.3 5.NR.4.3 Use place value understanding to round decimal numbers to the hundredths place. (DOK 1)

Round decimals to the nearest whole number.

1. 9,879.23

2. 89,797.45

3. 3,637.33

4. 378.9

5. 338.5

6. 30,484.434

7. 21.39

Round decimals to the nearest tenth.

8. 4.7894

9. 4,983.32

10. 8,497.45

11. 987,424.09

12. 43,768.284

13. 6.586

Round decimals to the nearest hundredth.

14. 489.75

15. 2.484

16. 44.4543

17. 4,848.345

18. .0495

19. 32,494.483

20. 45,009.993

Unit 6.4 5.NR.4.4 Solve problems involving addition and subtraction of decimal numbers to the hundredths place using a variety of strategies. (DOK 2)

You can use unit squares to add simple decimals. This is covered on the PowerPoint.

When adding decimals, please make sure that you align the decimals up on top of each other. This will help to keep the numbers being added in the right columns and context.

Ex. 1 Add 734.91 + 395.28

$$\begin{array}{r} 734.91 \\ +395.28 \\ \hline \mathbf{1130.19} \end{array}$$

Add the following decimals:

1. $\begin{array}{r} 844.2 \\ +339.2 \\ \hline \end{array}$

2. $\begin{array}{r} 892.7 \\ +273.23 \\ \hline \end{array}$

3. $\begin{array}{r} 621.854 \\ +345.678 \\ \hline \end{array}$

4. $\begin{array}{r} 1972.34 \\ +828.27 \\ \hline \end{array}$

5. $\begin{array}{r} 1945.97 \\ +234.03 \\ \hline \end{array}$

6. $\begin{array}{r} 348.63 \\ +23.87 \\ \hline \end{array}$

7. $\begin{array}{r} 54.987 \\ +6.303 \\ \hline \end{array}$

8. $\begin{array}{r} 6.045 \\ +5.09 \\ \hline \end{array}$

Subtract the following decimals:

Ex. 2 743.87 - 732.32

Again, align the decimals in order to subtract the two numbers.

$$\begin{array}{r} 743.87 \\ -732.32 \\ \hline \end{array} \qquad \begin{array}{r} 743.87 \\ -732.32 \\ \hline 11.55 \end{array}$$

9. $\begin{array}{r} 765.32 \\ -345.92 \\ \hline \end{array}$ **10.** $\begin{array}{r} 834.4 \\ -382.2 \\ \hline \end{array}$ **11.** $\begin{array}{r} 432.8 \\ -389.1 \\ \hline \end{array}$ **12.** $\begin{array}{r} 765.31 \\ -62.3 \\ \hline \end{array}$

13. $\begin{array}{r} 853.32 \\ -65.16 \\ \hline \end{array}$ **14.** $\begin{array}{r} 975.58 \\ -23.50 \\ \hline \end{array}$ **15.** $\begin{array}{r} 540.65 \\ -20.43 \\ \hline \end{array}$ **16.** $\begin{array}{r} 1952.87 \\ -872.36 \\ \hline \end{array}$

Align the decimals and add or subtract.

17. 837.34 + 283.5

18. 483.25 + 238.75

19. 763.21 - 32.01

20. 764.97 - 321.32

Decimal Quiz

1. Which of the following represents the number 7.853 in expanded form?

A) 7 + 8 + 5 + 3

B) 7 + 0.8 + 0.05 + 0.003

C) 7 + 0.3 + 0.5 + 0.8

D) 7 + 0.53 + 0.01

2. Compare the numbers 9.876 and 9.678. Which is greater?

A) 9.876

B) 9.678

C) They are equal

D) Cannot be determined

3. Round the number 9.782 to the nearest tenth.

A) 9.7

B) 9.8

C) 9.2

D) 9.4

4. Subtract 985.32-345.3

A) 83.765

B) 650.3

C) 2.041

D) 640.02

5. Which of the following numbers is ordered from least to greatest?

A) 0.356, 0.365, 0.346

B) 0.465, 0.346, 0.256

C) 0.446, 0.456, 0.465

D) 0.378, 0.386, 0.402

6. Round 22.654 to the nearest hundredth.

A) 22.645

B) 22.64

C) 22.65

D) 22.37

7. Add 7.34 + 2.14

A) 9.42

B) 9.44

C) 9.46

D) 9.48

8. Which expression correctly shows 0.579 in standard form?

A) 579

B) 0.0579

C) 0.579

D) 5.79

9. What is the difference? 7.35-4.21

A) 3.81

B) 3.65

C) 3.14

D) 4.52

10. If a graph shows the average monthly snowfall as 2.15 inches, 2.85 inches, and 3.7 inches for three months, what is the snow total for the three months?

A) 7.8 inches

B) 8.56 inches

C) 8.355 inches

D) 8.7 inches

Unit 6.5 5.MDR.7.2 Ask questions and answer them based on gathered information, observations, and appropriate graphical displays to solve problems in everyday life. (DOK 2)

Ex. 1 When Aaron and his dad went to the dealership, Aaron saw a newer convertible Mustang. The manager mentioned that the Mustang would be about $36,000. If the dad put $6,000 down, how much would need to be paid for 4 years? To simplify, we will use 0 % APR, the annual percentage rate.

Ans, $36,000 - $6,000 = $30,000. We know that Aaron's dad was considering paying off the car in 4 years. There are 48 months in 4 years. So, we divide the remaining $30,000 by 48 months. Therefore the remaining monthly payments would be $625 per month.

Ex. 2 Howard High School is selling tickets to a drama performance to raise money for new equipment. If the tickets sold on day 1 were $725.00, on day 2 were $870.00, and on day 3 were $478.50, what is the total amount of money raised by the school?

Ans, we'd add the totals raised each day. $725 + $870 + $478.5 = $2,073.5

Solve the following:

1. Sadie had a set of puppies. Their weights were 2.25 lbs, 3.18 lbs, 2.76 lbs, and 2.89 lbs. What are the total weights of the puppies?

2. Autumn has a couple of kittens. If one kitten weighs 2.23 lbs, and the total weight of the kittens is 4.77 lbs, what is the weight of the other kitten?

3. Taylor is packing a box to send to his cousin. The paint brushes' weight is 2.25 grams, the water colors' weight is 5.78 grams, and the canvasses' weights are 5.74 grams and 5.63 grams. What is the total weight of the art supplies that Taylor is planning to send to his cousin?

4. Taylor decides to send his niece twice as many art supplies than his cousin. What is the total weight of the art supplies that he will send to his niece?

5. Connor is estimating how much he needs to save to buy a Honda to get to school. The car he is looking at costs $12,000. He plans to put $2,000 down. How much will he plan to pay each month if he plans to pay off the car in 4 years at 0% APR?

Spiral Quiz

Explain your steps as you solve it.

John, David, Michael, and Alex went camping for 2, 3, 5, and 6 weeks. David wasn't there for three weeks. John stayed the longest, and Alex stayed longer than David. How many weeks did each stay at camp?

At a fundraiser for the homeless, the local pastor said he would give a prize to whoever is able to bring in over $500. Anna sells her tickets to 48 people, and each ticket was $11.65. Will she win the prize?

Unit 6 Review

5. NR.4.1 Read and write decimal numbers to the thousandths place using base-ten numerals, written in standard form and expanded form.

5.NR.4.2 Represent, compare, and order decimal numbers to the thousandths place based on the meanings of the digits in each place, using >, =, and < symbols to record the results of comparisons.

5.NR.4.3 Use place value understanding to round decimal numbers to the hundredths place.

5.NR.4.4 Solve problems involving addition and subtraction of decimal numbers to the hundredths place using a variety of strategies.

5.MDR.7.2 Ask questions and answer them based on gathered information, observations, and appropriate graphical displays to solve problems relevant to everyday life.

Read and write decimal numbers to the thousandths place.

1. Two thousand three hundred fifty-two and thirty-three hundredths.

2. Ten thousand four hundred eighty-three and twenty-six hundredths.

3. Seven hundred forty-six and eight hundred fifty-two thousandths.

4. Two thousand three hundred forty-five and six hundred seventy-eight thousandths.

5. Eight thousand five hundred ninety-seven and four hundred sixty-two thousandths.

Change the following decimals into words.

6. 4,893.24

7. 392.345

8. 1,982.697

9. 267.348

10. 0.592

11. 18,235.27

Compare decimals using >, <, =.

12. 483.23 ____ 485.92

13. 753.02 ____ 3,739.21

14. 3,903.43 ____ 3,842.384

15. 49,093.38 ____ 29,384.332

16. 702.98 ____ 384.92

17. 848.3 ____ 892.32

18. 6,554.78 ____ 673.92

Order decimals from least to greatest.

19. 483.93 38.387 493.02

20. 438.45 3,890.32 392.32

21. 382.34 38.392 38,349.2

22. 8,343.2 382.212 29,303.22

23. 76.545 677.43 646.43

24. 2,738.33 2,201.23 382.32

Add the decimals.

25. 456.82 + 3,948.31

26. 4,849.33 + 8,382.47

27. 523.29 + 739.01

28. 354.06 + 218.75

29. 84.027 + 347.83

30. 7,850.35 + 437.801

31. 3,782.23 + 1,829.23

32. 2,939.12 + 182.22

Subtract the decimals.

33. 827.32 - 392.39

34. 7,322.98 - 5,837.77

35. 9,848.43 - 783.45

36. 38,292.87 - 3,829.56

37. 5,788.54 - 2,765.54

38. 44,593.38 - 4,323.17

39. 635.876 - 632.345

40. 4,392.82 - 2,629.34

41. Trevor wants to go to the mall to buy an outfit for the dance. He has $500 that he can spend. He buys a shirt for $87.50 and a pair of pants for $75.76. He finds a pair of shoes to buy for $129.53. Does he have enough money to buy his outfit?

42. Carly is budgeting her money for her bills during the month. Her mortgage is $958.75, her electric bill is around $175.25, and her water bill is about $35.50 a month. The car insurance is $175, and the car payment is $299.20 a month. If she makes $2,345.80 a month, does she have enough money to pay her bills and have some to save and/or invest?

Unit 7

Exploring Geometry and the Coordinate Plane

Unit 7 Vocabulary

These are the vocabulary words to be discussed in this unit. Try to have fun with them and discuss them with your learning partners and stakeholders, i.e., parents, students, coaches.

Attribute: The characteristic of an object, i.e., size, color.

Axes: Two number lines that intersect at right angles on the origin of the coordinate plane.

Coordinate: A pair of numbers used to determine the position of a point located on the coordinate plane.

Coordinate plane (Cartesian plane): A 2-dimensional surface where points are plotted using ordered pairs.

First quadrant: Top right section of the coordinate plane.

Ordered pair: A pair of numbers or expressions written in the order (x,y).

Origin: The center of the coordinate plane. Point where the x- and y- axes intersect.

Polygon: A closed shape with straight sides.

X-coordinate: The first number in an ordered pair.

Y-coordinate: The second number in an ordered pair.

Create a story with pictures using these vocabulary words. Share with friends.

Unit 7.1 5.PAR.6.1 Generate Two Numerical Patterns (DOK 2)

5.PAR.6.1 Students will generate two numerical patterns and identify relationships between the corresponding terms in those patterns.

We see patterns on a daily basis. Numerical patterns can be seen in Art, Math, Finances, and even Nature. Look at a pine cone and notice the pattern. Those are called fractals. The spiral of a shell is called a Golden Ratio.

Example 1: You decide to work a summer job and earn $10/hour. Or $80 for Mon, $80 for Tue, $80 for Wed, $80 for Thu, and $80 for Fri. This is a numerical pattern for earning money during the weekdays. What would be the pattern? $10x where x begins at 1.

Example 2: Consider the following pattern: 3, 6, 9, 12, 15

a) State the rule to find the next number. Ans. 3x, where x begins at 1.

b) What would be the 6th number in the pattern? Ans. The 6th number will be 18.

Example 3: Create a numerical pattern where 3 is added to the next number, beginning with the number 5. Ans. 5, 8, 11, 14, 17, _

Example 4: Consider the following pattern: 2, 8, 32, 128...

a) State the rule to find the next number.

b) Calculate the next two numbers in the pattern.

Exercises:

Complete the patterns by filling in the missing numbers:

1. 5, __, 15, __, 25, __, 35

2. 3, __, 9, 12, 15, __, 21

3. 4, __, 10, 13, __, 16

4. 20, 17, __, 11, __

5. 7, 13, __, 25, __

Consider the following numerical patterns: State the rule for the patterns.

6. 5, 8, 11, 14, 17

7. 2, 6, 10, 14, 18

8. 1, 8, 15, 22, 29

9. 4, 8, 12, 16, 20

10. 7,14,21,28,35

What are the next two numbers of the following numerical patterns:

11. 2, 6, 10, 14

12. 3,9, 15, 21

13. 5, 12, 19, 26,

Let's look at numerical patterns:

Example 1: Given Pattern 1: Starting with 3, add 1. From Pattern 2: 5(x)-4

Step 1: Let's look at the first 5 terms generated by each pattern.

Step 2: Let's create a table detailing this information.

Step 3: We'll look for patterns to compare and contrast these relationships.

Term	Pattern 1 Start 3 +1	Pattern 2 5(x)-4	Difference
1	4	1	-3
2	5	6	1
3	6	11	5
4	7	16	9
5	8	21	13

Look at the following patterns, and use a table to find the relationships between the two patterns.

1. Pattern 1: Start at 5 and add 2 each time.

Pattern 2: Start at 3 and add 5 each time.

Term	Pattern 1	Pattern 2	Difference

2. Pattern 1: Start at 2 and add 3 each time.

Pattern 2: Start at 1 and add 4 each time.

Term	Pattern 1	Pattern 2	Difference

3. Pattern 1: Start at 15 and subtract 2 each time.

Pattern 2: Start at 17 and subtract 3 each time.

Term	Pattern 1	Pattern 2	Difference

4. Pattern 1: Start at 16 and subtract 2 each time.

Pattern 2: Start at 24 and subtract 3 each time.

Term	Pattern 1	Pattern 2	Difference

5. Pattern 1: Start at 5 and multiply by 1 each time.

Pattern 2: Start at 3 and multiply by 2 each time.

Term	Pattern 1	Pattern 2	Difference

6. Pattern 1: 3(x)+2

Pattern 2: 2(x)+5

Term	Pattern 1	Pattern 2	Difference

Next, look at numbers patterns to create a rule for the pattern.

Example 2: 5,10,15,20,25 Starting with 5 add 5.

Look at the numerical pattern and create a rule.

7. 6,10,14,18,22

8. 4,7,10,13,15

9. 25,23,21,19,17

10. 5,9,13,17,21

Unit 7.2 5.PAR.6.2 Represent problems by plotting ordered pairs and explain coordinate values of points in the first quadrant of the coordinate plane. (DOK 1)

Imagine that you have decided to take a weekend vacation. You wouldn't mind some company to visit for everyone to experience fishing and/or hiking. Now, imagine trying to explain to your friend how to get somewhere in an unfamiliar place. Wouldn't you know it? Your cell service is spotty. You decide to text him and include a photo of a map detailing where you are at. One way you can describe your location is by using an ordered pair. Your friend can use this location to find you on the map.

Now, we are going to try to simplify this as much as possible. One thing we are going to look at is graphing an ordered pair in the coordinate plane. The coordinate plane has four quadrants. They are numbered counter clockwise about the origin. The ordered pairs that are graphed on this plane are in the form of (x,y). And yes, we could add a z, but that would make it three-dimensional, and we will save that thought for a later time.

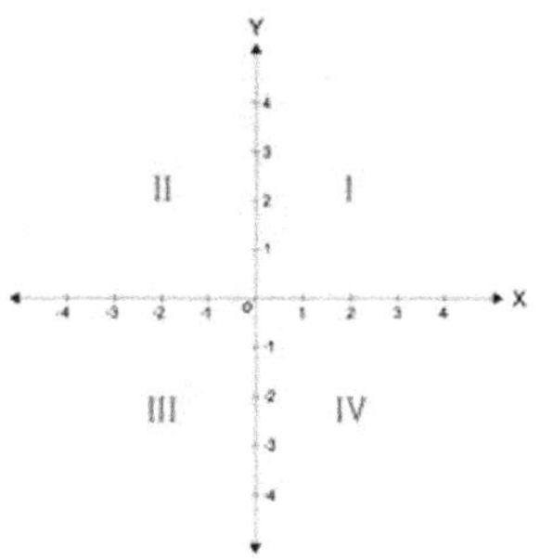

The Coordinate Plane: (aka The Cartesian Plane) Notice that the x-axis runs horizontally and the y-axis runs vertically.

In order to make this explanation as clear as possible, we are going to concentrate on Quadrant I. Why just Quad 1 ? Well, we want this to be a positive experience.

Example 1: Let's say Josh wants to graph point A(4,5) on the coordinate plane below. He would begin at the origin (0,0). The x-coordinate moves the point to the left or right; in this example, we are going 4 units to the right. The y-coordinate moves the point to either up or down. In this example, we are going to move the point up 5 units. In summary, we are plotting point A by moving 4 units to the right and 5 units up.

Example 2: Plot B(3,2) Answer: You'd start at the origin, move right 3 units, and up 2 units.

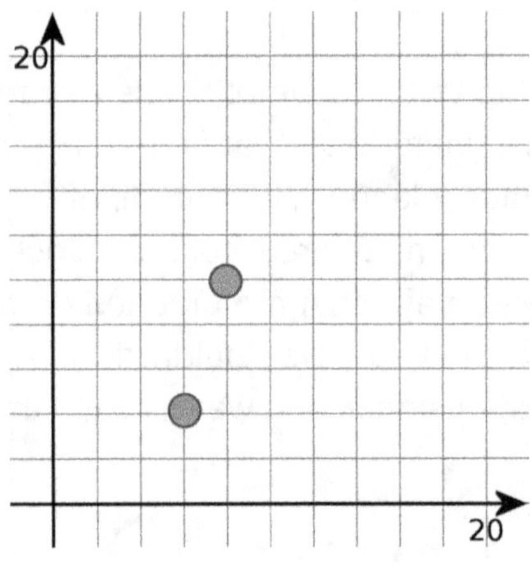

Example 3: Points can be located on a coordinate plane. If you are asked to state the ordered pair of the point. Start at the origin, move over to the right the number of units needed, and move up the number of units needed to be on the point. Write the coordinate as your ordered pair. **Ans.** (5,5)

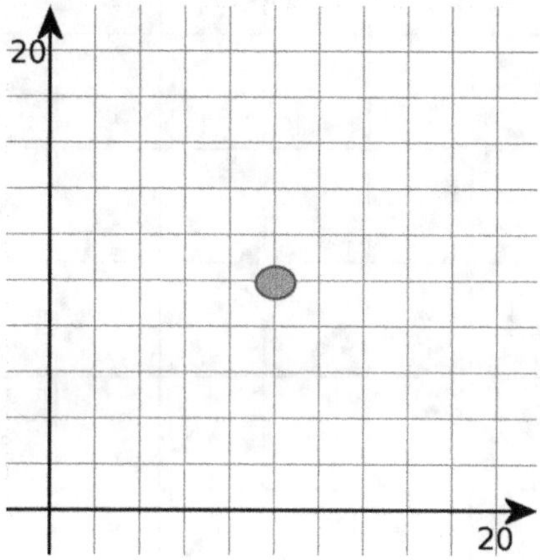

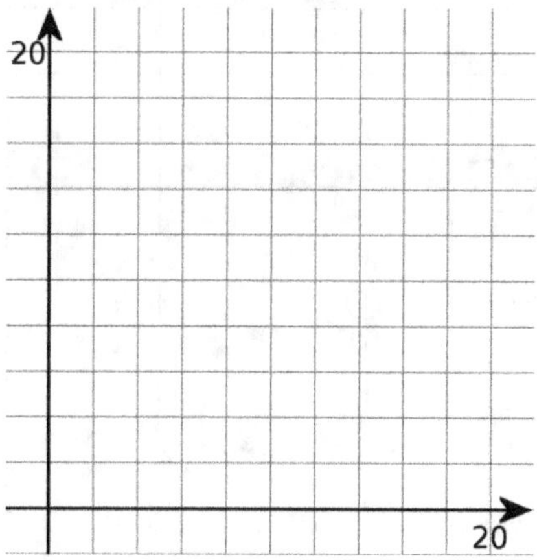

Given the following ordered pairs, plot the points on the coordinate plane.

1. A (2,2) **2.** B (4,6) **3.** C (1,7) **4.** D (0,10)

5. E (9,0) **6.** F (0,0) **7.** G (5,3) **8.** H (7,2)

The next part to look at is given the points on the coordinate plane: determine the ordered pair. State the order pairs of the points on the coordinate plane below.

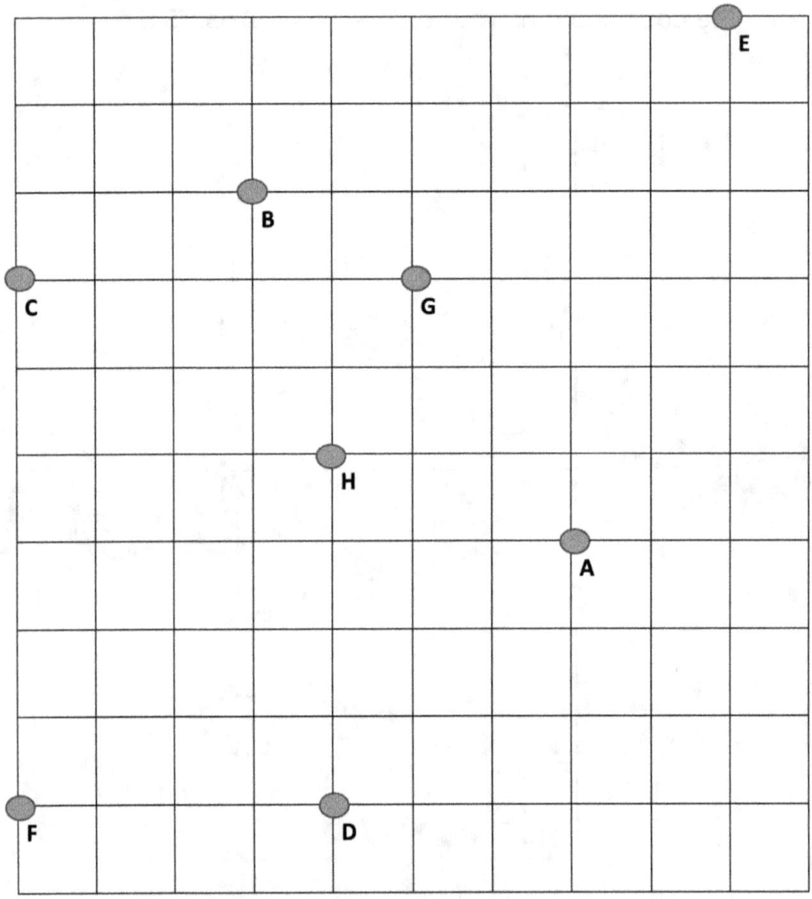

Record the ordered pairs of the points on the coordinate plane.

9. A **10.** B **11.** C **12.** D

13. E **14.** F **15.** G **16.** H

Unit 7.3 5.GSR.8.1 Classifying, Comparing, and Contrasting Polygons (DOK 2)

In this lesson, we will explore the fascinating world of polygons. Understanding the properties of polygons will help us classify, compare, and contrast them effectively. Polygons are everywhere around us, from the shapes of road signs to the structure of buildings. Let's delve in and learn more about these multi-sided figures.

Understanding Polygons

Polygons are two-dimensional shapes with straight sides. The simplest polygon is a triangle, which has three sides. As the number of sides increases, the complexity and uniqueness of the polygon increases.

Polygons can be classified based on several properties:

Sides and Angles: The number of sides and angles a polygon has. For example, a triangle has three sides.

Regular and Irregular: A regular polygon has all sides and angles equal, while an irregular polygon does not.

Convex and Concave: A convex polygon has no angles pointing inward, whereas a concave polygon has at least one angle that points inward.

Example 1: Triangle

Properties: Three sides and three angles.

Can be classified into different types, such as equilateral (all sides equal), isosceles (two sides equal), and scalene (no sides equal).

Example 2: Quadrilateral

Properties: Four sides and four angles.

Includes squares, rectangles, and trapezoids.

Squares and rectangles are regular quadrilaterals, while trapezoids are irregular.

Concave and Convex: Concave curves inwards, and Convex curves outwards

Use the properties discussed to classify, compare, and contrast the following polygons. Determine whether they are regular or irregular, and if they are convex or concave.

1. An octagon with all sides equal.

2. A trapezoid with two sides longer than the others.

3. A quadrilateral with opposite sides equal.

4. A triangle with one angle larger than 90 degrees.

5. A triangle with no equal sides or angles.

6. A pentagon with all sides equal.

7. A decagon with all angles equal.

8. A nonagon with one angle smaller than the others.

9. A dodecagon with all equal angles.

10. A square.

11. A right-angled triangle.

12. A rectangle.

13. A rhombus with all sides equal.

14. A parallelogram with unequal adjacent sides.

15. A kite with two pairs of equal sides.

16. An equilateral triangle.

17. A scalene triangle.

18. A trapezoid with one pair of parallel sides.

19. A concave hexagon.

20. A convex octagon.

21. A regular hexagon.

22. A star-shaped polygon.

23. A quadrilateral with two right angles.

24. A pentagon with one obtuse angle.

25. A heptagon with all sides of different lengths.

Unit 7.4 5.GSR.8.2 Determine, through exploration and investigation, that attributes belonging to a category of two-dimensional figures also belong to all subcategories of that category. (DOK 2)

Consider two-dimensional figures. Two-dimensional figures have length and width. Let's attempt to group them into categories and subcategories based on their attributes. Let's consider how attributes of a category apply to its subcategories: it's crucial for grasping geometric concepts.

Example 1: Quadrilaterals

Category: Quadrilaterals

Subcategories: Parallelograms, Rectangles, Squares

Attribute: A quadrilateral is a polygon with four sides.

Application to Subcategories:

Parallelograms have opposite sides that are parallel.

Rectangles are parallelograms with four right angles.

Squares are rectangles with all sides equal in length.

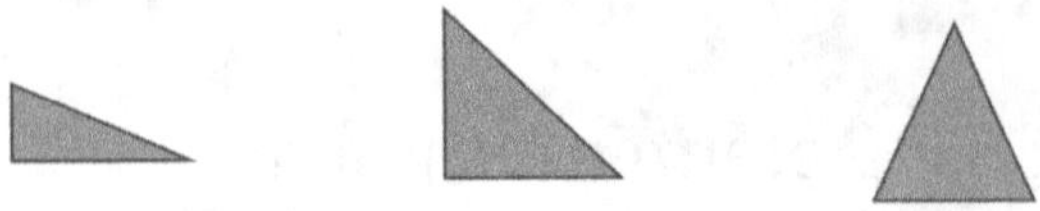

Example 2: Triangles

Category: Triangles

Subcategories: Equilateral, Isosceles, Scalene

Attribute: A triangle is a polygon with three sides.

Application to Subcategories:

Equilateral triangles have all sides equal.

Isosceles triangles have at least two sides equal.

Scalene triangles have all sides of different lengths.

Example 3: Circles

Category: Circles

Subcategories: Concentric Circles, Sectors, Semicircles

Attribute: A circle is a set of points equidistant from a center point.

Application to Subcategories:

Concentric circles share the same center but have different radii.

Sectors are portions of a circle bounded by two radii and an arc.

Semi-circles are half of a circle.

Questions

1. Name a common attribute in all squares.

2. How does the attribute of having three sides apply to triangles?

3. Is a rectangle a square, or a square a rectangle? Explain.

4. What is the difference between a parallelogram and other quadrilaterals?

5. How does an isosceles triangle differ from an equilateral triangle?

6. What is a common attribute of a rectangle?

7. What geometric property is shared by all quadrilaterals?

8. Are circles actually a pair of semicircles?

9. What is a sector in a circle?

10. How many pairs of parallel sides does a parallelogram have?

11. Can a rhombus also be a kite? Explain.

12. How do you classify a triangle with three different angles?

13. What is in common between a circle and a semicircle?

14. Is a circle a polygon?

15. Do radii and circumference have something in common?

16. Name a polygon that is symmetric.

17. What attributes do a square and a rhombus have in common?

18. By definition, a polygon is a what? Many angles or many gons?

19. How does a rectangle differ from a rhombus?

20. What is the sum of the interior angles of a square?

21. What is the sum of the interior angles of a triangle?

Use this space to create your own attribute of two different polygons:

Test your attribute: Does it hold true for the polygons you chose? Others?

Unit 7 Review

5. PAR.6.1 Generate two numerical patterns using two given rules. Identify apparent relationships between corresponding terms by completing a table.

5.PAR.6.2 Represent problems by plotting ordered pairs and explain coordinate values of points in the first quadrant of the coordinate plane.

5.GSR.8.1 Classify, compare, and contrast polygons based on properties.

5.GSR.8.2 Determine, through exploration and investigation, that attributes belonging to a category of two-dimensional figures also belong to all subcategories of that category.

Numerical patterns

1. Rule 1: Starting with 4, add 3

Rule 2: Starting with 1, add 4

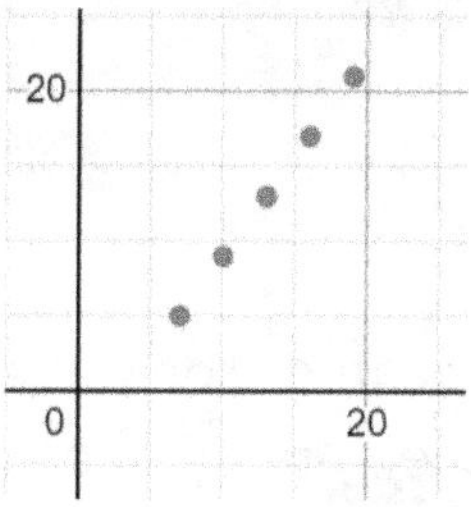

Term	Pattern 1	Pattern 2
1	4+3 = 7	1+4 = 5
2	7+3 = 10	5+4 = 9
3	10+3 = 13	9+4 = 13
4	13+3 = 16	13+4 = 17
5	16+3 = 19	17+4 = 21

2. Rule 1: Starting with 1, add 5

Rule 2: Starting with 3, add 3

Term	Pattern 1	Pattern 2
1	1+5 = 6	6
2	6+5 = 11	9
3	11+5 = 16	12
4	16+5 = 21	15
5	21+5 = 26	18

Determine the relationship between pattern 1 and pattern 2

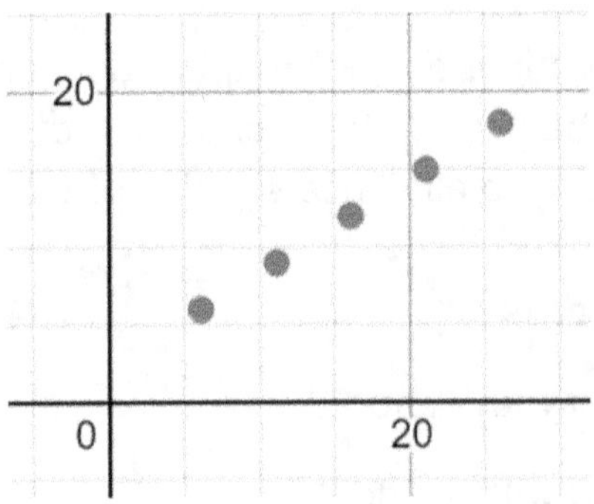

3. Origin is where the x and y axis cross.

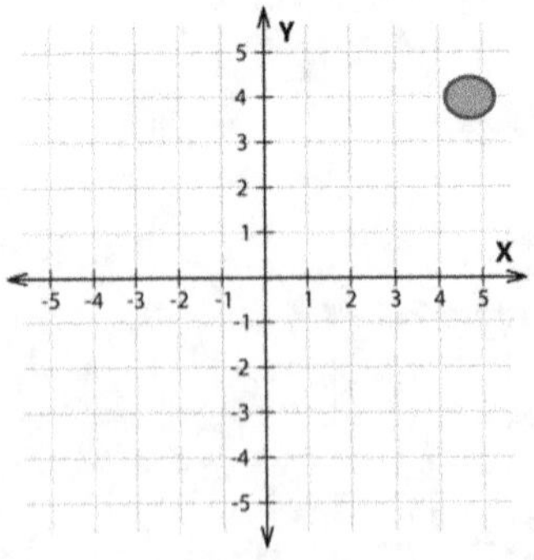

The points located on the coordinate plane are mapped using ordered pairs, (x,y).

Ex. (5,4)

4. Graph the following points on the coordinate plane.

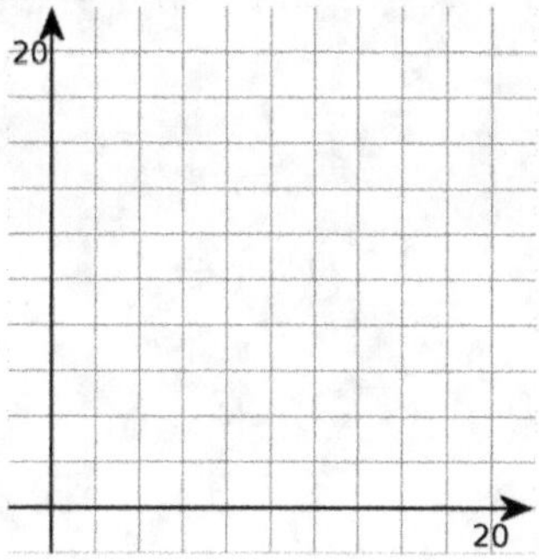

A (4,6)

B (5,0)

C (2,8)

D (0,18)

E (4,4)

5. Classify and compare polygons based on properties.

Circles - no angles, 360 degrees

Triangle - 3 angles, 3 sides, 180 degrees

Quadrilaterals - 4 angles, 4 sides, 360 degrees

Compare and contrast - a scalene triangle has 3 different angles and 3 different sides. An isosceles triangle has 2 equal angles and one different angle, and two equal length sides and one different length side.

Both are triangles. Both have 180 degrees.

Think of two other shapes to compare and contrast.

5th Grade Post-test

1. Evaluate the expression: 5.NR.5.1 (Dok 2)

Given: $5b-6c+(3a+2b)^2$. Given: a = 2, b = 3, and c = 1

2. Simplify: 3(4-2) 5.NR.5.1 (Dok 1)

3. What kind of property is used? 5.NR.5.1 (Dok 1)

A. Identity B. Commutative C. Distributive

4. Solve for x: 8x = 48 5.NR.2.1 (Dok 1)

A. x = 5 B. x = 6 C. x = 6 D. x = 5

5. Find the volume of a cube with 5in edge lengths. 5.GSR.8.3 (Dok 1)

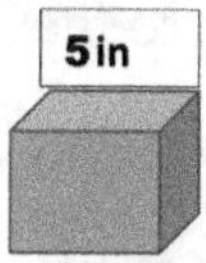

A. $64in^3$ B. $75in^3$ C. $81in^3$ D. $125in^3$

6. The number of cubes that filled a rectangular prism is 32. If the height of the rectangular prism is 4 units, find the area of the base of the rectangular prism. 5.GSR.8.3

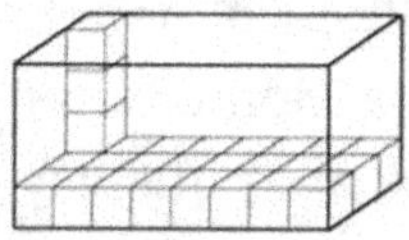

A. 4 $units^2$ B. 8 $units^2$ C. 9 $units^2$ D. 6 $units^2$

7. Evaluate 7×5 - (4+2). 5.NR.5.1 (Dok 1)

A. 24 B. 32 C. 36 D. 29

8. Find the volume of a rectangular prism having a length of 2 ft a width of 2.5 ft, and a height of 12 ft. 5.DSR.8.4 (Dok 2)

A. 20 $units^3$ B. 30 $units^3$ C. 50 $units^3$ D. 60 $units^3$

9. Diego and Conner had bought a pizza to share. 5.NR.3.3 (Dok 2)

Diego had eaten 1/8 of it before Conner got a chance to eat 2/5 of it.

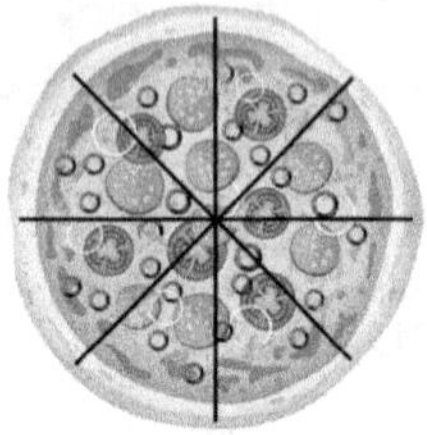

How much pizza do they have to eat for the next day?

10. Jason was saving money for a new bike for his birthday.

If the bike costs $120, and he has 1/5 of it, how much money does he have left to save? 5.NR.3.4 (Dok 2)

A. $96 B. $95 C. $100 D. $110

11. Marco was explaining to his sister that the following shapes are similar. A rhombus, a triangle, and a square. How are they similar? 5.GSR.8.1 (Dok 1)

A. They are shapes B. They have angles

C. They have sides. D. They are two-dimensional

12. Kyle and Elise were cleaning their rooms before their grandparents arrived. They decided that it would be a good idea to display the model cars and horse figures on the shelves in their rooms. One horse is the size of three model cars. If a toy horse is a foot long, and a model car is 4 inches long, about how many of each would fit on two three-foot shelves? 5.MDR.7.2 (Dok 3)

A. 6 toy horses or 18 toy cars B. 1 toy horse or 3 toy cars

C. A and B D. Cannot be determined

13. Given the number: 259,648. 5.NR.1.1 (Dok 1)

Which number is located in the ten thousands place?

A. 2 B. 5 C. 9 D. 6

14. Add decimals: 9,753.2 and 85.34. 5.NR.4.4 (Dok 1)

A. 9838.54 B 8938.53 C. 8938.54 D. 9848.52

15. Jordan sees an ad for a new online game on the internet. The ad says that there will be prizes given for the 1st through 3rd place winners. The first prize is $20; the second prize is half of that, and the third prize is a fourth of the first-place prize. If Jordan considers for a moment that there are about a thousand contestants and twice as many who clicked to see the ad, should he try to win the 1st place prize of skipping the ad? 5.NR.3.1 (Dok 2)

A. Yes B. No C. Sample too small D. Cannot be determined

16. Mountains in Colorado are 15×10^3 feet above sea level. Mountains in Georgia are around 20×10^2. How much higher are the mountains in Colorado compared to the mountains in Georgia? Express your answer in powers of 10. 5.NR.1.2 (Dok 2)

A. Nope B. 10^4 C. 13×10^3 D. 13×10^2

17. Subtract: . $\frac{3}{4} - \frac{1}{5}$.

5.NR.3.3 (Dok 2)

A. 3/20 B. 4/5 C. 11/20 D. 17/20

18. Karla's class is going to Lookout Mountain to see the Incline Railway. If there are thirty-three students, two teachers, and three chaperons, how many activity vans are needed if one van holds twelve people and a driver?

A. 1 B. 2 C. 3 D. 4 5.NR.3.1 (Dok 2)

19. Angel's parents bought her a used car for college. It had 76,500 miles on it when they purchased it. After graduating college, she had 265,700 miles listed on the odometer. How many miles did Angel drive her car once she got it?

A. 189300 B. 191300 C. 187100 D. 189200 5.MDR.7.2 (Dok 2)

20. Drake's dad has a yard of free-roaming chickens. If there are three times as many hens as roosters, how many chickens would be in his yard if Drake had counted 8 roosters? What is the ratio of roosters to hens and the ratio of hens to chickens, minus the cat? 5.NR.3.4 (Dok 2)

A. 24:8 B. 3:1 C. 4:3 D. 5:2

21. Josh and Jacob are at a fishing tournament. The smallest fish can be 2 and a half pounds. Josh could easily qualify since his sunfish weighs 5 and two-thirds pounds. What is the difference in the weights? 5.NR.3.3 (Dok 2)

A. 3 1/4 B. 2 3/4 C. 3 1/6 D. 2 1/8

22. A recreation department received a shipment of 50 footballs, 30 soccer balls, and 60 basketballs. Which expression shows the amount of sports equipment received.

A. 100 B. 90 C. 140 D. 120 5.NR.4.4 (Dok 1)

23. Which two expressions are equivalent to 4x + 8y?

A. 2(x+3y) B. -2(x+3y) C. 4(x+2y) D. -4(-x-2y) 5.NR.1 (Dok 2)

24. What is the rule for the expression? 24, 30, 36, 42, 48

A. Add 2 B. Add 4 C. Add 6 D. Add 8 5.PAR.6.1 (Dok 2)

25. How many inches are in a mile? 5.MDR.7.2 (Dok 1)

A. 5,280 B. 26,740 C. 46,780 D. 63,360

26. Divide 3,875 by 25. 5.NR.2.2 (Dok 2)

A. 135 B. 150 C. 155 D. 160

27. Which of the following are improper fractions? 5.NR.3.1 (Dok 2)

A. 1/4 B. 2/3 C. 3/2 D. 3/5

28. Change 27/2 to a mixed number. 5.NR.3.1 (Dok 1)

A. 12 1/2 B. 13 1/2 C. 14 1/2 D. 15 1/2

29. Multiply 2/3 by 4. 5.NR.3.4 (Dok 2)

A. 2 2/3 B. 2 1/6 C. 3 1/4 D. 4 3/4

30. Which engine has more horsepower, 289 or 357?

A. 289 B. 197 C. 48 D. 357 5.MDR.7.1 (Dok 1)

31. Kayla is confused. Her grandmother sent her to the store to get 4 eggs for a meal. How many dozen eggs would she need to buy for her grandmother? 5.MDR.7.4 (Dok 2)

A. 2.5 B. 3 C. 3.5 D. 4

32. Justin's parents have convinced their son that he needs a job for the summer. His mom said that she would give him $10 for the chores that he completes. Not to be outdone, his dad said that he would give him double the amount of chores he completes, starting with a nickel. After a month of completing 20 chores, which offer would have benefited Justin the most? 5.PAR.6.2 (Dok 2)

A. $150 & $300 B. $120 & $300

C. $100 & $300 D. $200 & $52,428.5

33. Given the following fractions, order them from least to greatest. 1/3 3/4 2/5 1/5 5.NR.3.2 (Dok 2)

34. Given the following decimals, order them from least to greatest. 2.3 5.2 3.6 1.8. 5.NR.4.2 (Dok 1)

35. Graph the following points on a coordinate plane.

A. (3,6) B. (5,6) 5.PAR.6.2 (Dok 1)

C. (2,8) D. (1,0). Which point is on an axis?

36. Which points are in Quadrant I? 5.PAR.6.2 (Dok 1)

A. (1,5) B. (-2,2) C. (-3,-2) D. (4,-3).

37. Which points are on the origin? 5.PAR.6.2 (Dok 1)

A. (0,4) B. (0,3) C. (0,0) D. (-4,0).

38. Multiply 2,890 by 34. 5.NR.2.1 (Dok 1)

39. Divide 2,970 by 15. 5.NR.2.2 (Dok 1)

40. Round the number to the nearest hundredths place:

9,837.245 5.NR.4.3 (Dok 1)

41. Classify the polygon. A circle: 5.GSR.8.1

A. No angles B. No sides C. 3 sides D. 3 angles

Teacher's Guide

1. 5th Grade Standards Georgia

5.NR.1.1 Explain that in a multi-digit number, a digit in one place represents 10 times as much as it represents in the place to its right, and 1/10 of what it represents in the place to its left.

5.NR.1.2 Explain patterns in the placement of digits when multiplied or divided by a power of 10. Use whole-number exponents to denote powers of 10, up to 10^3.

5.NR.2 Multiply and divide multi-digit whole numbers to solve relevant mathematical problems.

5.NR.2.1 Fluently multiply multi-digit (up to 3-digit by 2-digit) whole numbers to solve authentic problems.

5.NR.2.2 Fluently divide multi-digit whole numbers (up to 4-digit dividends and 2-digit divisors no greater than 25) to solve practical problems.

5.NR.3 Describe fractions and perform operations with fractions to solve relevant mathematical problems using part-whole strategies and visual models.

5.NR.3.1 Explain the meaning of a fraction as division of the numerator by the denominator ($a/b = a \div b$). Solve problems involving division of whole numbers leading to answers in the form of fractions or mixed numbers. ($a/b =$)

5.NR.3.2 Compare and order up to three fractions with different numerators and/or different denominators by flexibly using a variety of tools and strategies.

5.NR.3.3 Model and solve problems involving addition and subtraction of fractions and mixed numbers with unlike denominators.

5.NR.3.4 Model and solve problems involving multiplication of a fraction and a whole number.

5.NR.3.5 Explain why multiplying a whole number by a fraction greater than one results in a product greater than the whole number, and why multiplying a whole number by a fraction less than one results in a product less than the whole number and multiplying a whole number by a fraction equal to one results in a product equal to the whole number.

5.NR.3.6 Model and solve problems involving division of a unit fraction by a whole number and a whole number by a unit fraction.

5.NR.4 Read, write, and compare decimal numbers to the thousandths place, and round and perform operations with decimal numbers to the hundredths place to solve relevant mathematical problems.

5.NR.4.1 Read and write decimal numbers to the thousandths place using base-ten numerals written in standard form and expanded form.

5.NR.4.2 Represent, compare, and order decimal numbers to the thousandths place based on the meanings of the digits in each place, using >, =, and < symbols to record the results of comparisons.

5.NR.4.3 Use place value understanding to round decimal numbers to the hundredths place.

5.NR.4.4 Solve problems involving addition and subtraction of decimal numbers to the hundredths place using a variety of strategies.

5.NR.5 Write, interpret, and evaluate numerical expressions within authentic problems.

5.NR.5.1 Write, interpret, and evaluate simple numerical expressions involving whole numbers with or without grouping symbols to represent actual situations.

5.PAR.6 Solve relevant problems by creating and analyzing numerical patterns using the given rule(s).

5.PAR.6.1 Generate two numerical patterns using two given rules. Identify apparent relationships between corresponding terms by completing a table.

5.PAR.6.2 Represent problems by plotting ordered pairs and explain coordinate values of points in the first quadrant of the coordinate plane.

5.MDR.7 Solve problems involving customary measurements, metric measurements, and time and analyze graphical displays of data to answer relevant questions.

5.MDR.7.1 Explore realistic problems involving different units of measurement, including distance, mass, weight, volume, and time.

5.MDR.7.2 Ask questions and answer them based on gathered information, observations, and appropriate graphical displays to solve problems relevant to everyday life.

5.MDR.7.3 Convert among units within the metric system and then apply these conversions to solve multistep, practical problems.

5.MDR.7.4 Convert among units within relative sizes of measurement units within the customary measurement system.

5.GSR.8 Examine properties of polygons and rectangular prisms, classify polygons by their properties, and discover volume of right rectangular prisms.

5.GSR.8.1 Classify, compare, and contrast polygons based on properties.

5.GSR.8.2 Determine, through exploration and investigation, that attributes belonging to a category of two-dimensional figures also belong to all subcategories of that category.

5.GSR.8.3 Investigate volume of right rectangular prisms by packing them with unit cubes without gaps or overlaps. Then, determine the total volume to solve problems.

5.GSR.8.4 Discover and explain how the volume of a right rectangular prism can be found by multiplying the area of the base times the height to solve authentic mathematical problems.

Unit 1 Standard: Investigating Volume of Solid Figures

5.GSR.8: Examine properties of polygons and rectangular prisms, classify polygons by their properties, and discover volume of right rectangular prisms.

5.NR.5: Write, interpret, and evaluate numerical expressions within real-life problems.

Grade: 5th	Subject: Math

Domain: Investigating Volume of Solid Figures	Big Idea: Geometric & Spatial Reasoning and Numerical Reasoning	Standard Topic: Volume
5.GSR.8.3 Investigate volume of right rectangular prisms by packing them with unit cubes without gaps or overlaps. Then, determine the total volume to solve problems. **5.GSR.8.4 Discover and explain how the volume of a right rectangular prism can be found by multiplying the area of the base times the height to solve real-life, mathematical problems.** **5.NR.5.1 Write, interpret, and evaluate simple numerical expressions involving whole numbers with or without grouping symbols to represent real-life situations.**		

Know Now: What students should know before class.	Do Now: What students should be able to do.
Students should know about area and multiplication before class.	**Students should be able to solve volume problems by multiplying the area of the base times height.**

Learning Targets
Students should be able to pack unit blocks into rectangular prisms and/or use unit cubes to create rectangular prisms. Students should be able to investigate that the number of cubes in the base is the area of the base. Understand that the area of the base times the height is the volume of the rectangular prism. Students should be able to generalize that the area of the base times the height is the volume of the rectangular prism.

Instructional Strategies
Warm-up: In two sentences, explain how to find the number of cubic units of a rectangular prism. Review the concept of polygons. Have students discuss what a polygon is. Have students give counter-examples to discuss what is not a polygon. Ask students to explain ways to find the volume of a polygon. Review finding volume by counting the number of unit cubes packed into a rectangular prism. Review that the area of the base times the height is the volume of the rectangular prism. Create numerical expressions that equal the volume of the rectangular prism.

Evidence of Learning
Students are able to find volume by counting unit cubes. Students can extend this into finding the volume of a rectangular prism by multiplying the base times the height. **Students can translate this idea to more than one rectangular prism.** **Justify that there are several different numerical expressions that can equal to the volume of a rectangular prism.**

5.GSR.8.3 Standard: Unit 1: 5.GSR.8: Examine properties of polygons and rectangular prisms, classify polygons by their properties, and discover volume of right rectangular prisms.

Grade: 5th	Subject: Math

Domain: Investigating Volume of Solid Figures	Big Idea: Geometric & Spatial Reasoning and Numerical Reasoning	Standard Topic: Volume
Standard Expectation: 5.GSR.8.3 Investigate volume of right rectangular prisms by packing them with unit cubes without gaps or overlaps. Then, determine the total volume to solve problems.		

Know Now: What students should know before class.	Do Now: What students should be able to do.
Students should know about area and multiplication before class.	Students should be able to solve volume problems by counting unit cubes.

Learning Targets
Investigate the volume of right rectangular prisms by packing them with unit cubes without gaps or overlaps. Then, determine the total volume to solve problems. Given a volume of a number, create a rectangular prism with unit cubes.

Instructional Strategies
Warm-up: In two sentences, describe what a rectangular prism is. Elijah wants to pack toy cubes to send to his cousins. The shipping carton is a rectangular prism that is 12in long, 9in wide, and 6in high. The cubes he wants to ship are 1 cubic inch. How many cubes will fit at the base of the shipping box? With unit cubes, create this scenario. Count the number of cubes at the base of the box, then continue until the box dimensions are filled. Was your answer correct? Yes or No? Determine the total volume of a prism by using a hands-on investigation of packing cubes. Students count the total number of unit cubes in the rectangular prism.

Evidence of Learning
Students are able to find volume by counting unit cubes. Students can translate this idea to more than one rectangular prism.

5.GSR.8.4 Lesson Plan

Standard: Unit 1:5.GSR.8: Examine properties of polygons and rectangular prisms, classify polygons by their properties, and discover volume of right rectangular prisms.

Grade: 5th	Subject: Math

Domain: Investigating Volume of Solid Figures	Big Idea: Geometric & Spatial Reasoning and Numerical Reasoning	Standard Topic: Volume
5.GSR.8.4 Discover and explain how the volume of a right rectangular prism can be found by multiplying the area of the base times the height to solve real-life, mathematical problems.		

Know Now: What students should know before class.	Do Now: What students should be able to do.
Students should know about area and multiplication before class.	**Students should be able to solve volume problems by multiplying the area of the base times height.**

Learning Targets
Students should be able to investigate that the number of cubes in the base is the area of the base. Understand that the area of the base times the height is the volume of the rectangular prism.

Instructional Strategies
Warm-up: In two sentences, explain how to find the number of cubic units of a rectangular prism. **Review the concept of polygons. Have students discuss what a polygon is. Have students give counter-examples to discuss what is not a polygon. Ask students to explain ways to find the volume of a polygon.** **Review finding volume by counting the number of unit cubes packed into a rectangular prism.**

Evidence of Learning
Students are able to find volume by counting unit cubes. Students can extend this into finding the volume of a rectangular prism by multiplying the base times the height. **Students can translate this idea to more than one rectangular prism.**

5.NR.5.1 Lesson Plan

Standard: Unit 1: 5.NR.5: Write, interpret, and evaluate numerical expressions within real-life problems.

Grade: 5th	Subject: Math	
Domain: Investigating Volume of Solid Figures	**Big Idea: Geometric & Spatial Reasoning and Numerical Reasoning**	**Standard Topic: Volume**
Standard Expectation: 5.NR.5.1 Write, interpret, and evaluate simple numerical expressions involving whole numbers with or without grouping symbols to represent real-life situations.		

Know Now: What students should know before class.	**Do Now: What students should be able to do.**
Students should know about area and multiplication before class.	**Students should be able to write, interpret, and evaluate simple numerical expressions involving whole numbers with or without grouping symbols to represent real-life situations.**

Instructional Strategies
Objective: **Warm-up: In two sentences, describe what a numerical expression is.** **Review the concept of rectangular prisms. Have students review what a rectangular prism is. Have students look at a rectangular prism built with unit cubes.** **Activity: Ask students to count the number of unit cubes on the base of the rectangular prism. Use 5.NR.1 PowerPoint to introduce simplifying numerical expressions. Ask students to create numerical expressions that are equivalent to the volume of the rectangular prism.**

Evidence of Learning
Assessment: Tell your neighbor the informal definition of volume. Give an example of finding the volume of a right triangular prism using a numerical expression to do so.

Unit 1 Test

Volume and Unit Cubes

1. The area of the base of a rectangular prism is 50, and there are 4 layers on top of the bottom layer. How many unit cubes were used to pack into the rectangular prism?

A. 100

B. 150

C. 200

D. 250

2. The area of the base is 30, with an additional layer added on top. What is the total volume of the rectangular prism?

A. 10 $units^3$

B. 20 $units^3$

C. 30 $units^3$

D. 60 $units^3$

3. Josh measured the base of the rectangular prism as 9 units in length and 8 units in width. With 2 additional layers, how many unit cubes are in the rectangular prism?

A. 132

B. 72

C. 164

D. 216

4. Conner used 80 unit cubes to create a rectangular prism with 4 layers. What could be the area of the base?

A. 10 $units^2$

B. 15 $units^2$

C. 20 $units^2$

D. 25 $units^2$

5. A rectangular prism has a base area of 12 square units. What is the volume if there are 3 layers?

A. 20 $units^3$

B. 36 $units^3$

C. 48 $units^3$

D. 24 $units^3$

6. What is the volume of a unit cube with side lengths of 4 units?

A. 10 $units^3$

B. 40 $units^3$

C. 50 $units^3$

D. 60 $units^3$

7. If the volume of a rectangular prism is 60 $units^3$, what would be the volume if this prism's dimensions are tripled?

A. 50 $units^3$

B. 120 $units^3$

C. 180 $units^3$

D. 200 $units^3$

Numeric Expressions

8. What is the sum of 24 and 48?

A. 60

B. 72

C. 84

D. 96

9. Calculate the product of 16 and 39.

A. 524

B. 624

C. 704

D. 784

10. What is 4 plus the sum of 21 and 19?

A. 44

B. 42

C. 34

D. 40

11. Find the difference of 62 and 9.

A. 51

B. 53

C. 55

D. 57

12. Evaluate 15(2 + 8).

A. 100

B. 150

C. 160

D. 170

13. Calculate 5(7) - 4(5).

A. 15

B. 25

C. 35

D. 45

14. Find the result of 10×32 + 7×8.

A. 320

B. 336

C. 376

D. 416

15. Evaluate 10x + 3y when x = 2 and y = 5.

A. 25

B. 35

C. 45

D. 55

Additional Questions

16. If the base area of a rectangular prism is doubled, how does the volume change if the height remains constant?

A. Volume is halved

B. Volume is doubled

C. Volume remains the same

D. Volume is quadrupled

17. What is the volume of a cube with a side length of 5 units?

A. 25

B. 75

C. 100

D. 125

18. A rectangular prism has a volume of 240 cubic units. If the area of the base is 40 square units, how many layers are there?

A. 4

B. 5

C. 6

D. 7

19. How many unit cubes are needed to fill a rectangular prism with a base area of 15 and height of 8?

A. 120

B. 130

C. 140

D. 150

20. What is the total number of unit cubes in a prism with a base area of 9 and 3 layers?

A. 18

B. 27

C. 36

D. 45

21. A rectangular prism has a base area of 22 and 4 layers. What is the volume?

A. 88 $units^3$

B. 98 $units^3$

C. 108 $units^3$

D. 118 $units^3$

22. What is the result of the expression 3(10) + 2(20)?

A. 40

B. 50

C. 60

D. 70

23. Calculate 8(4) - 2(3).

A. 20

B. 22

C. 24

D. 26

24. Evaluate 7x - 5y when x = 3 and y = 2.

A. 11

B. 13

C. 15

D. 17

25. What is the volume of a rectangular prism with dimensions 3, 4, and 5 units?

A. 60 $units^3$

B. 70 $units^3$

C. 80 $units^3$

D. 90 $units^3$

Unit Plan: Unit 2

Unit 2 Standard: Unit 2: Building Conceptual Understanding of Place Value Using Measurement and Data Reasoning.

5.NR.1: Use place value understanding to solve real-life mathematical problems.

5.MDR.7: Solve problems involving customary measurements, metric measurements, and time and analyze graphical displays of data to answer relevant questions.

Grade: 5th	Subject: Math

Domain: Building Conceptual Understanding of Place Value Using Measurement and Data Reasoning	Big Idea: Building Conceptual Understanding of Place Value Using Measurement and Data Reasoning	Standard Topic: Place Value

5.NR.1.1 Explain that in a multi-digit number, a digit in one place represents 10 times as much as it represents in the place to its right, and 1/10 of what it represents in the place to its left.
5.NR.1.2 Explain patterns in the placement of digits when multiplied or divided by a power of 10. Use whole-number exponents to denote powers of 10, up to 10^3.
5.MDR.7.3 Convert among units within the metric system and then apply these conversions to solve multi-step, real life problems.
5.MDR.7.4 Convert among units within relative sizes of measurement units within the customary measurement system.
5.MDR.7.1 Explore real-life problems involving different units of measurement, including distance, weight, volume, and time.
5.MDR.7.2 Ask questions and answer them based on gathered information, observations, and appropriate graphical displays to solve problems relevant to everyday life.

Learning Targets
Solve problems by representing mathematical relationships between quantities using mathematical expressions and equations. **Investigate the effects of multiplying whole numbers by powers of 10.** **Convert units within the same measurement system by multiplying or dividing using conversion factors.** **Use line plots to display a data set of measurements, which includes fractions.** **Solve problems based on data displayed in a line plot.** **Select appropriate units, strategies, and tools for solving problems that involve estimating and measuring volume.** **Communicate precisely by engaging in discussion about their reasoning using appropriate mathematical language.**

Instructional Strategies
Warm-up: In two sentences, explain how to find the number of cubic Identify a number in the ones place, tens place, etc. Explain that a multi-digit number, a digit in one place represents 10 times as much as it represents in the place to its right and 1/10 of what it represents in the place to its left. Explain patterns in the placement of digits when multiplied or divided by a power of 10. Use whole-number exponents to denote powers of 10, up to 10^3. Convert among units within the metric system and then apply these conversions to solve multi-step, real life problems. Convert among units within relative sizes of measurement units within the customary measurement system. Explore real-life problems involving different units of measurement, including distance, weight, volume, and time. Ask questions and answer them based on gathered information, observations, and appropriate graphical displays to solve problems relevant to everyday life.

Evidence of Learning
State the informal definition of place value. Give an example of a multi-digit number and correctly label the correct place value of the digits. Given a number, explain that a number is ten times or 1/10 the value of a digit next to it. Convert a given value into different measurements in customary systems and metric systems.

Unit 2 Standard: Unit 2: Building Conceptual Understanding of Place Value Using Measurement and Data Reasoning

5.NR.1: Use place value understanding to solve real-life mathematical problems.

5.MDR.7: Solve problems involving customary measurements, metric measurements, and time and analyze graphical displays of data to answer relevant questions.

Grade: 5th	Subject: Math

Domain: Place Value	Big Idea: Numerical Reasoning and Measurement and Data Reasoning	Standard Topic: Dividing fractions
Standard Expectation: 5.NR.1.1 Explain that in a multi-digit number, a digit in one place represents 10 times as much as it represents in the place to its right, and 1/10 of what it represents in the place to its left.		

Know Now: What students should know before class.	Do Now: What students should be able to do.
Students should know how to identify place value.	Students should be able to explain that a multi-digit number in one digit is 10 times or 1/10 the number on its left or right.

Learning Targets
Students will use base ten blocks to visually create numbers and show how the place value system is used to efficiently multiply and divide powers of 10.

Instructional Strategies
Caleb is looking at a number line. He sees the numbers 2 and 20. Draw the two numbers on the number line. Looking at the numbers, what is similar and what is different about the two numbers? How do you know? **Review: Place value.** **Activity: Use the Unit 2 Place Value PowerPoint to discuss Place Value. Use a number line if necessary.**
Assessment: Tell your neighbor the similarity and difference of the numbers 50 and 500.

5.NR.1.2 Unit 2 Standard: Unit 2: Building Conceptual Understanding of Place Value Using Measurement and Data Reasoning

5.NR.1: Use place value understanding to solve real-life mathematical problems.

5.MDR.7: Solve problems involving customary measurements, metric measurements, and time and analyze graphical displays of data to answer relevant questions.

Grade: 5th	Subject: Math

Domain: Place Value	Big Idea: Numerical Reasoning and Measurement and Data Reasoning	Standard Topic: Place Value
Standard Expectation: 5.NR.1.2 Explain patterns in the placement of digits when multiplied or divided by a power of 10. Use whole-number exponents to denote powers of 10, up to 10^3.		

Know Now: What students should know before class.	Do Now: What students should be able to do.
Students should know how to identify place value, multiply, and divide by 10.	**Students should be able to explain that a multi-digit number in one digit is 10 times or 1/10 the number on its left or right.**

Learning Targets
Students will use base ten blocks to visually create numbers and show how the place value system is used to efficiently multiply and divide powers of 10.

Instructional Strategies
Warm-up: In two sentences, describe what place value is. Give an example. **Given the number 1,375, identify the number in the hundreds place.** **Given the number 28,634, identify the number in the thousands place.** **Review the place value of the digits.** **Activity: Use 5.NR.1 PowerPoint to introduce simplifying numerical expressions. Use 5.NR.1.2 assignment.**
Assessment: Tell your neighbor the similarity and difference of the numbers 500 and 5000.

6.MDR.7.3 Convert among units within the metric system and then apply these conversions to solve multi-step, real life problems.

Unit 2 Standard: Unit 2: Building Conceptual Understanding of Place Value Using Measurement and Data Reasoning

5.NR.1: Use place value understanding to solve real-life mathematical problems.

5.MDR.7: Solve problems involving customary measurements, metric measurements, and time and analyze graphical displays of data to answer relevant questions.

Grade: 5th	Subject: Math

Domain: Place Value	Big Idea: Numerical Reasoning and Measurement and Data Reasoning	Standard Topic: Place Value
Standard Expectation: 5.MDR.7.3 Convert among units within the metric system and then apply these conversions to solve multi-step, real life problems.		

Know Now: What students should know before class.	Do Now: What students should be able to do.
Students should know how to identify place value, multiply, and divide by 10.	Students should be able to explain how to convert among units within the metric system. Students should be able to apply what the learn about moving decimals by multiplication and division.

Learning Targets
Students will use base ten blocks to visually create numbers and show how the place value system is used to efficiently multiply and divide powers of 10. Students can use a metric table or graphic organizer to help them keep track of their conversions.

Instructional Strategies
Warm-up: In two sentences, describe how you know that there are 200 cm in 1 meter. **Given the measurement 1,375 mm, convert it to cm. Convert it to m.** **Review the place value of the digits.** **Activity: Use 5.NR.1 PowerPoint to introduce simplifying numerical expressions. Use 5.NR.1.2 assignment.**
Assessment: Tell your neighbor how you know that there are 5000ml in 5 liters.

Unit 2 Standard: Unit 2: Building Conceptual Understanding of Place Value Using Measurement and Data Reasoning

5.NR.1: Use place value understanding to solve real-life mathematical problems.

5.MDR.7: Solve problems involving customary measurements, metric measurements, and time and analyze graphical displays of data to answer relevant questions.

Grade: 5th	Subject: Math	
Domain: Place Value	**Big Idea: Numerical Reasoning and Measurement and Data Reasoning**	**Standard Topic: Place Value**
Standard Expectation: 5.MDR.7.4 Convert among units within relative sizes of measurement units within the customary measurement system.		

Know Now: What students should know before class.	Do Now: What students should be able to do.
Students should know how to identify place value, multiply, and divide by whole numbers.	Students should be able to explain how to convert among units within the customary system. Students should be able to apply what they learn about multiplication and division.

Learning Targets
Students will use customary measurements to create conversions. Students can use a graphic organizer to help them keep track of their conversions. However, it is good for them to be able to calculate them instead of lots of customary measurements given to them.

Instructional Strategies
Warm-up: In two sentences, explain what you know about the yard measurement. **Given the measurement 30 yards, convert it to feet. Convert it to inches.** **Review the place value of the digits.** **Activity: Use 5.NR.1 PowerPoint to introduce simplifying numerical expressions. Use 5.NR.1.2 assignment.**
Assessment: Tell your neighbor how you know that 5000 inches is more or less than a mile.

5.MDR.7.1 Explore real-life problems involving different units of measurement, including distance, weight, volume, and time.

Unit 2 Standard: Unit 2: Building Conceptual Understanding of Place Value Using Measurement and Data Reasoning

5.NR.1: Use place value understanding to solve real-life mathematical problems.

5.MDR.7: Solve problems involving customary measurements, metric measurements, and time and analyze graphical displays of data to answer relevant questions.

Grade: 5th	Subject: Math

Domain: Place Value	Big Idea: Numerical Reasoning and Measurement and Data Reasoning	Standard Topic: Place Value
Standard Expectation: 5.MDR.7.1 Explore real-life problems involving different units of measurement, including distance, weight, volume, and time.		

Know Now: What students should know before class.	Do Now: What students should be able to do.
Students should be able to draw on prior knowledge to identify measurements of distance, weight, volume, and time.	Students should be able to explain how to convert among units within the customary system. Students should be able to apply what the learn by using multiplication and division. They should know relative sizes and be able to know when to use them.

Learning Targets
Students will apply place value reasoning in order to convert between customary units of measure, focusing on lengths.

Instructional Strategies
Objective: **Warm-up: In two sentences, explain what you know about changing 2 miles to feet.** **Cody wants to know how many inches are in the length of a football field. Explain to him how to find the length of a football field in inches. Explain how you know the correct number of inches. (Hint: work backwards.)**
Review the place value of the digits. **Activity: Use the Unit 2 PowerPoint to introduce the customary system. The assignment Units of Measurement is helpful for students to begin to understand the customary system and be able to change between units of measurement.**
Assessment: Tell your neighbor how you know that number of inches in 2 miles.

Unit 2 Standard: 5.MDR.7.2: Building Conceptual Understanding of Place Value Using Measurement and Data Reasoning

5.NR.1: Use place value understanding to solve real-life mathematical problems.

5.MDR.7: Solve problems involving customary measurements, metric measurements, and time and analyze graphical displays of data to answer relevant questions.

Grade: 5th	Subject: Math

Domain: Place Value	Big Idea: Numerical Reasoning and Measurement and Data Reasoning	Standard Topic: Place Value
Standard Expectation: 5.MDR.7.2 Ask questions and answer them based on gathered information, observations, and appropriate graphical displays to solve problems relevant to everyday life.		

Learning Targets
Students will apply place value reasoning in order to convert between customary units of measure, focusing on lengths.

Instructional Strategies
Objective: **Warm-up: In two sentences, explain what you know about changing 2 miles to feet.** **Cody wants to know how many inches are in the length of a basketball court. Explain to him how to find the length of a basketball court in inches. Explain how you know the correct number of inches.** **Review the place value of the digits.** **Activity: Use the Unit 2 PowerPoint to introduce the customary system. The assignment Ask Questions is helpful for students to begin to understand the customary system and be able to change between units of measurement.**

Assessment: Tell your neighbor the difference between the metric and the customary system. Give an example and explain the difference between the systems.

Unit 2 Test

Number Place Value and Comparison

1. Consider the hundreds place and the tens place. Compare the hundreds place to the tens place.

A. The hundreds place is 10x greater than the tens place.

B. The hundreds place is 1/10 greater than the tens place.

2. Consider the thousands place and the hundreds place. Compare the thousands place to the hundreds place.

A. The thousands place is 10x greater than the ones place.

B. The hundreds place is 1/10 greater than the thousands place.

3. The value of the 5 in the number 85,732 is ____ times the value of the 5 in the number 78,534.

A. 1

B. 10

C. 100

D. 1/10

4. The value of the 3 in the number 63,571 is ___ times the value of the 3 in the number 3,765.

A. 1

B. 10

C. 100

D. 1/10

5. Write the following number in standard form: 90,000 + 8,000 + 400 + 20 + 3.

A. 98432

B. 98421

C. 98342

D. 98423

6. Write the following number in expanded form: 94723.

A. 90,000 + 4,000 + 700 + 20 + 3

B. 90,000 + 4,000 + 700 + 3

C. 94,000 + 7,000 + 20 + 3

D. 90,000 + 4,000 + 720 + 3

Multiplication and Division

7. Multiply 3645 x 100.

A. 364500

B. 3645000

C. 238200

D. 365400

8. Multiply 83 x 10,000.

A. 83000

B. 8300

C. 830

D. 830000

9. When multiplying 64 by 1,000, how many powers of 10 is 64 multiplied by?

A. 1

B. 2

C. 3

D. 4

10. Multiply 53×10^2.

A. 3500

B. 530

C. 53000

D. 5300

11. Divide 96,000 by 10^3.

A. 96

B. 960

C. 690

D. 6900

12. Divide 7500000 by 10^4.

A. 75000

B. 57000

C. 750

D. 7500

Writing Numbers as a Factor of 10

13. Write 831,000,000 as a factor times the power of 10.

A. 831×10^4

B. 831×10^5

C. 831×10^6

D. 831×1

14. Write 670,000 as a factor times the power of 10.

A. 67×10^4

B. 67×10^5

C. 67×10^6

D. 67×10^7

15. Write 920,000,000 as a factor times the power of 10.

A. 92×10^5

B. 92×10^6

C. 92×10^7

D. 92×10^8

Metric System Conversion

16. 1800 m = ____ km

A. 1.8

B. 0.18

C. 18

D. 180

17. 203 m = ____ cm

A. 20300

B. 2030

C. 203

D. 20.3

18. 831 ml = ____ L

A. 8.31

B. 0.831

C. 83.1

D. 8310

19. 1240 g = ____ kg

A. 1.24

B. 12.4

C. 0.124

D. 124

Customary System Conversion

20. 3 miles = ____ in

A. 190080

B. 95040

C. 63360

D. 5280

21. 48 ft = _____ yd

A. 16

B. 144

C. 12

D. 24

22. 196 in = _____ ft

A. 16.33

B. 16.5

C. 14

D. 16

Word Problems

23. A rectangle has a length of 25 in and a width of 9 in. What is the area of the rectangle?

A. 150 ft^2

B. 175 in^2

C. 200 ft^2

D. 225 in^2

24. If one box contains 48 crayons, how many dozen crayons are in a carton of 60 boxes?

A. 220

B. 230

C. 240

D. 250

25. A baker wants to sell brownies for the festival. If he bakes 180 brownies, how many dozen brownies can he sell?

A. 10

B. 15

C. 20

D. 30

26. Callie makes $1430 per week. What is the place value for the 4 in the number?

A. 10

B. 20

C. 50

D. 100

Unit 3 Standard: Building Conceptual Understanding of Multiplication and Division with Whole Numbers

5.NR.2: Multiply and divide multi-digit whole numbers to solve real-life, mathematical problems.

5.NR.5: Write, interpret, and evaluate numerical expressions within real-life problems.

5.MDR.7: Solve problems involving customary measurements, metric measurements, and time and analyze graphical displays of data to answer relevant questions.

Grade: 5th	Subject: Math

Domain: Building Conceptual Understanding of Multiplication and Division with Whole Numbers	Big Idea: Building Conceptual Understanding of Place Value Using Measurement and Data Reasoning	Standard Topic: Whole Numbers
5.NR.2.1 Fluently multiply multi-digit (up to 3-digit by 2-digit) whole numbers to solve real-life problems. **5.NR.5.1 Write, interpret, and evaluate simple numerical expressions involving whole numbers with or without grouping symbols to represent real-life situations.** **5.NR.2.2 Fluently divide multi-digit whole numbers (up to 4-digit dividends and 2-digit divisors no greater than 25) to solve practical problems.** **5.MDR.7.2 Ask questions and answer them based on gathered information, observations, and appropriate graphical displays to solve problems relevant to everyday life.**		

Know Now: What students should know before class.	Do Now: What students should be able to do.
Students should know about building on their conceptual understanding of place value, properties of operations, multiplication and division. Some should have experience with area diagrams.	Students should be able to use steps for multiplication and division. Students should be able to collect, organize, and interpret data.

Learning Targets
Students should be able to multiply and divide multi-digit whole numbers to solve problems. Students should be able to write, interpret, and evaluate numerical expressions in the real world. Ask questions on gathered information, observation, and graphical displays involving customary units of measure.

Instructional Strategies
Multiply and divide whole numbers using a variety of techniques based on place value, part to whole, and properties of operations. Students understand different ways to write a product. Students may use an algorithm or an area model. Students write equivalent division expressions. Students will discuss the partial quotient strategy to be used in division problems. Understand that place value is important when multiplying and dividing whole numbers. Students will write, evaluate, and interpret mathematical expressions used in everyday life.

Multiplication is repeated adding, and division is repeated subtraction.

Students will explore using the order of operations to simplify mathematical expressions.

Ask questions and answer them based on gathered information, observations, and appropriate graphical displays to solve problems relevant to everyday life.

Evidence of Learning
Multiplication and division are explained and computed correctly. Use of order of operations to simplify mathematical expressions.

Unit 3 Standard: 5.NR.2: Multiply and divide multi-digit whole numbers to solve real-life, mathematical problems.

5.NR.5: Write, interpret, and evaluate numerical expressions within real-life problems.

5.MDR.7: Solve problems involving customary measurements, metric measurements, and time and analyze graphical displays of data to answer relevant questions.

Grade: 5th	Subject: Math

Domain: Building Conceptual Understanding of Multiplication and Division with Whole Numbers	Big Idea: Building Conceptual Understanding of Place Value Using Measurement and Data Reasoning	Standard Topic: Whole Numbers
Standard Expectation: **5.NR.2.1 Fluently multiply multi-digit (up to 3-digit by 2-digit) whole numbers to solve real-life problems.**		

Know Now: What students should know before class.	Do Now: What students should be able to do.
Students should be able to draw on prior knowledge of different techniques to multiply and divide whole numbers.	Students should be able to explain how to convert among units within the customary system and the metric system. Students should be able to apply what the learn by using multiplication and division. They should know relative sizes and be able to know when to use them.

Learning Targets
Students will apply place value reasoning in order to convert between customary units of measure, focusing on lengths.

Instructional Strategies
Objective: **Warm-up: In two sentences, explain what you know about multiplying a three-digit number by a two-digit number.** **Raven multiplies the number 745 by 32. She gets the number 21900. Is she correct?** **Review the place value of the digits.** **Activity: Use the Unit 3 PowerPoint to introduce or review multiplication. Review multiplying by single digits, then double digits. Work up to multiplying by a 3-digit by a 2-digit number. The assignment Making Sense of Multiplication is helpful for students to begin to understand multiplying by multi-digit numbers.**

Assessment **Ask a partner to create a multiplication problem. Try to work it out using calculation and an area model.**

5.NR.2.2 Fluently divide multi-digit whole numbers (up to 4-digit dividends and 2-digit divisors no greater than 25) to solve real-life problems.

5.NR.5.1 Write, interpret, and evaluate simple numerical expressions involving whole numbers with or without grouping symbols to represent real-life situations.

Unit 3 Standard: 5.NR.2: Multiply and divide multi-digit whole numbers to solve real-life, mathematical problems.

5.NR.5: Write, interpret, and evaluate numerical expressions within real-life problems.

5.MDR.7: Solve problems involving customary measurements, metric measurements, and time and analyze graphical displays of data to answer relevant questions.

Grade: 5th	**Subject: Math**

Domain: Building Conceptual Understanding of Multiplication and Division with Whole Numbers	**Big Idea: Building Conceptual Understanding of Place Value Using Measurement and Data Reasoning**	**Standard Topic: Whole Numbers**
Standard Expectation: 5.NR.5.1 Write, interpret, and evaluate simple numerical expressions involving whole numbers with or without grouping symbols to represent real-life situations.		

Know Now: What students should know before class.	**Do Now: What students should be able to do.**
Students should be able to draw on prior knowledge of different techniques to multiply and divide whole numbers.	**Students should be able to apply what the learn by using multiplication and division. They should know relative sizes and be able to know when to use them.**

Learning Targets
Students will apply numerical reasoning of simplifying expressions.

Instructional Strategies
Objective: **Warm-up: In two sentences, explain what a numerical expression is.** **Raven multiplies the number 45 by 32. She gets the number 2190. Is she correct?** **Review the Order of Operations.** **Activity: Use the Unit 3 PowerPoint to introduce or review multiplication. Review multiplying and dividing. Review addition and subtraction. The assignment Unit 3: 5.NR.5.1a is helpful for students to begin to understand simplifying expressions.**

Assessment: Tell your neighbor what an expression is. Give an example and explain how to simplify the expression.

Unit 3 Test

Determine the Property

1. $1 \times 25 = 25$

A. Associative

B. Commutative

C. Distributive

D. Identity

2. A number multiplied by zero

A. Distributive

B. Identity

C. Zero Property

D. Commutative

3. $2(5) = 5(2)$

A. Associative

B. Commutative

C. Distributive

D. Identity

4. $3 \times (3 \times 3) = (3 \times 3) \times 3$

A. Associative

B. Commutative

C. Distributive

D. Zero Property

5. $8 \times 1 = 8$

Which property is used?

A. Associative

B. Commutative

C. Distributive

D. Identity

6. $2(9 + 1) = 18 + 2$

Which property is used?

A. Associative

B. Commutative

C. Distributive

D. Zero Property

7. $4(2 \times 3) = (4 \times 2) \times 3$

Which property is used?

A. Associative

B. Commutative

C. Distributive

D. Identity

8. 5 + 0 = 5

Which property is used?

A. Associative

B. Commutative

C. Zero Property

D. Identity

9. 8(2+4) = 8(2) + 8(4)

Which property is used?

A. Associative

B. Commutative

C. Distributive

D. Identity

10. 2 + (3+ 4) = (2+3) + 4

Which property is used?

A. Associative

B. Commutative

C. Distributive

D. Identity

Multiply

11. 921 × 9

A. 8,289

B. 8,2891

C. 8,229

D. 8,191

12. 280 × 55

A. 15,400

B. 15,4000

C. 14,500

D. 15,000

13. 3,495 × 4

A. 13,980

B. 14,980

C. 13,890

D. 13,990

14. 2,018 × 43

A. 86,774

B. 86,7740

C. 87,774

D. 86,674

15. 600 × 40

A. 24,000

B. 24,0000

C. 23,000

D. 24,040

16. 295 × 27

A. 7,965

B. 7,9650

C. 7,975

D. 7,885

17. 325 × 64

A. 20,800

B. 20,8000

C. 20,900

D. 20,700

18. 7,860 × 25

A. 196,500

B. 196,5000

C. 196,600

D. 196,350

Divide

19. 455 ÷ 5

A. 91

B. 90

C. 92

D. 93

20. 2400 ÷ 25

A. 96

B. 95

C. 94

D. 97

21. 1412 ÷ 4

A. 353

B. 352

C. 354

D. 355

22. 8169 ÷ 3

A. 2,723

B. 2,733

C. 2,743

D. 2,753

23. 1,575 ÷ 15

A. 105

B. 103

C. 106

D. 104

24. 2,864 ÷ 2

A. 1,432

B. 1,422

C. 1,442

D. 1,452

25. 4,825 ÷ 25

A. 193

B. 192

C. 191

D. 194

Multiply using Part-Whole

26. Eric wants to find the area of a field. It is 65 yards in length and 19 yards in width. He chooses Part-whole thinking to help him multiply. Is his strategy right?

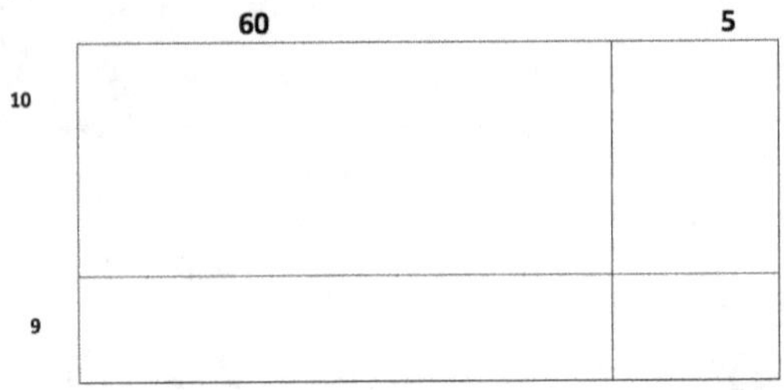

27. Create a part-whole for multiplying 45 by 15.

Illustrate your method.

Divide using Partial Quotient.

28. How many times can 5 go into 165?

We know that 5 × 20 = ?

We know that 5 × 13 = ?

Therefore, 165 divided by 5 = ?

29. Luis wants to know how many groups of 5 members can fit into the plane that holds 245.

He thinks: $5 \times 20 = 100$

$5 \times 20 = 100$

$7 \times 5 = 35$

Therefore, there should be 73 groups of 5 able to fit on the plane. Is he correct? Explain.

Unit 4: Building Fraction Understanding

5.NR.3: Describe fractions and perform operations with fractions to solve real-life mathematical problems using part-whole strategies and visual models.

5.MDR.7: Solve problems involving customary measurements, metric measurements, and time and analyze graphical displays of data to answer relevant questions.

Grade: 5th	**Subject: Math**

Domain: Fractions	**Big Idea: Numerical Reasoning and Measurement & Data Reasoning**	**Standard Topic: Fractions**
5.NR.3.2 Compare and order up to three fractions with different numerators and/or different denominators by flexibly using a variety of tools and strategies. **5.NR.3.3 Model and solve problems involving addition and subtraction of fractions and mixed numbers with unlike denominators.** **5.MDR.7.2 Ask questions and answer them based on gathered information, observations, and appropriate graphical displays to solve problems relevant to everyday life.**		

Know Now: What students should know before class.	Do Now: What students should be able to do.
Students should know about equivalent fractions.	**Students should be able to solve fraction problems by ordering fractions and finding equivalent fractions. Students should be able to add and subtract fractions and mixed numbers with unlike denominators. Ask questions based on gathered information, observations, and displays to solve relevant problems.**

Learning Targets
Student should be able to order fractions. They should be able to add fractions with unlike denominators. Generate equivalent fractions. **Students should be able to investigate lengths and measurements on the number line using fractions.** **Ask statistical questions involving fractions.**

Instructional Strategies
Compare fractions with like and unlike denominators. **Order fractions on a number line.** **Review the concept of estimation. Have students consider if a sum or a difference of fractions is reasonable.** **Ask and answer questions involving measurement and graphical displays of data.**

Evidence of Learning
Students are able to compare and order fractions with like and unlike denominators. Create their own set of ordered fractions. **Adding and subtracting fractions with like and unlike denominators.**

Unit 4 Lesson Plan

Unit 4 5.NR.3: Describe fractions and perform operations with fractions to solve real-life mathematical problems using part-whole strategies and visual models.

5.MDR.7: Solve problems involving customary measurements, metric measurements, and time and analyze graphical displays of data to answer relevant questions.

Grade: 5th	Subject: Math

Domain: Fractions	Big Idea: Numerical Reasoning and Measurement & Data Reasoning	Standard Topic: Fractions
Standard Expectation: 5.NR.3.2 Compare and order up to three fractions with different numerators and/or different denominators by flexibly using a variety of tools and strategies.		

Know Now: What students should know before class.	Do Now: What students should be able to do.
Students should be able to draw on prior knowledge of different techniques to multiply and divide whole numbers.	Students should be able to apply and extend what the learn by using multiplication and division. They should know relative sizes and be able to know when to use them.

Learning Targets
Students will fluently multiply and divide multi-digit whole numbers to solve practical problems. Students will gather information and data to be able to solve customary units of measure problems.

Instructional Strategies
Objective: **Warm-up: In two sentences, explain what you know about a fraction. Give an example.** **Jess orders three fractions 1/2, 2/3, and 3/5. Marc said that her ordering is incorrect. Is he correct? Justify your answer.** **Review: GCF and LCM.** **Activity: Use the Unit 3 PowerPoint for the assignment 5.NR.3.2 is helpful for students to begin to understand ordering fractions. We can order fractions by getting the common denominator and changing fractions to decimals.**

Assessment: Tell your neighbor two different ways to order fractions. Give an example and explain the difference between the two.

Unit 4 5.NR.3: Describe fractions and perform operations with fractions to solve real-life mathematical problems using part-whole strategies and visual models.

5.MDR.7: Solve problems involving customary measurements, metric measurements, and time and analyze graphical displays of data to answer relevant questions.

Grade: 5th	Subject: Math

Domain: Fractions	Big Idea: Numerical Reasoning and Measurement & Data Reasoning	Standard Topic: Fractions
Standard Expectation: 5.NR.3.3 Model and solve problems involving addition and subtraction of fractions and mixed numbers with unlike denominators.		

Know Now: What students should know before class.	Do Now: What students should be able to do.
Students should be able to draw on prior knowledge of different techniques to multiply and divide whole numbers.	Students should be able to apply and extend what the learn by using multiplication and division. They should know relative sizes and be able to know when to use them.

Learning Targets
Students will fluently multiply and divide multi-digit whole numbers to solve practical problems. Students will gather information and data to be able to solve customary units of measure problems.

Instructional Strategies
Objective: **Warm-up: In two sentences, explain what you know about adding and subtracting fractions. Give an example.** **Alex adds three fractions 1/2 and 3/5. Marc said that his adding of fractions should be 4/7. Is he correct? Justify your answer.** **Review: GCF and LCM.** **Activity: Use the Unit 3 PowerPoint for the assignment 5.NR.3.3 is helpful for students to begin to understand adding and subtracting fractions. We can add and subtract fractions by getting the common denominator and changing fractions to decimals.**

Assessment: Tell your neighbor two different ways to add and subtract fractions. Give an example and explain the difference between the two.

Unit 4 5.NR.3: Describe fractions and perform operations with fractions to solve real-life mathematical problems using part-whole strategies and visual models.

Unit 4 5.NR.3: Describe fractions and perform operations with fractions to solve real-life mathematical problems using part-whole strategies and visual models.

5.MDR.7: Solve problems involving customary measurements, metric measurements, and time and analyze graphical displays of data to answer relevant questions.

Grade: 5th	Subject: Math

Domain: Fractions	Big Idea: Numerical Reasoning and Measurement & Data Reasoning	Standard Topic: Fractions
Standard Expectation: **5.MDR.7.2 Ask questions and answer them based on gathered information, observations, and appropriate graphical displays to solve problems relevant to everyday life.**		

Know Now: What students should know before class.	Do Now: What students should be able to do.
Students should be able to draw on prior knowledge of different techniques to add and subtract fractions.	Students should be able to apply and extend what they learn by using multiplication and division. They should know how to find equivalent fractions in order to add and subtract fractions.

Learning Targets
Students will fluently add and subtract fractions with like and unlike denominators to solve practical problems. Students will gather information and data to be able to solve customary units of measure problems.

Instructional Strategies
Objective: Warm-up: In two sentences, explain what you know about adding and subtracting fractions. Give an example. Alex adds three fractions 1/2 and 3/5. Marc said that his adding of fractions should be 4/7. Is he correct? Justify your answer. Review: GCF and LCM. Activity: Use the Unit 3 PowerPoint for the assignment 5.NR.3.3 is helpful for students to begin to understand adding and subtracting fractions. We can add and subtract fractions by getting the common denominator and changing fractions to decimals.

Assessment: Tell your neighbor two different ways to add and subtract fractions. Give an example and explain the difference between the two.

Unit 4 Test

Compare Fractions Using >, <, and =

1. Which of the following symbols correctly compares the fractions 3/4 and 2/3?

A) >

B) <

C) =

D) Cannot be determined.

2. Choose the correct symbol to compare 5/8 and 3/4.

A) >

B) <

C) =

D) Can't wait for lunch.

3. Compare 7/10 and 7/9 using the appropriate symbol.

A) >

B) <

C) =

D) I don't think so.

4. Which symbol correctly compares 1/2 and 2/5?

A) >

B) <

C) =

D) Chevy or Ford?

5. Determine the correct comparison for 4/5 and 6/7.

A) >

B) <

C) =

D) Both.

6. Use a symbol to compare 1/3 and 1/4.

A) >

B) <

C) =

D) Cannot be determined.

Order Fractions from Least to Greatest

7. Order the fractions 1/2, 1/3, and 1/4 from least to greatest.

A) 1/2, 1/3, 1/4

B) 1/4, 1/3, 1/2

C) 1/3, 1/4, 1/2

D) 1/4 is the baby of the group.

8. Arrange the fractions 3/5, 2/3, and 1/2 in ascending order.

A) 1/2, 2/3, 3/5

B) 3/5, 1/2, 2/3

C) 1/2, 3/5, 2/3

D) Too many fractions.

9. Put the fractions 5/6, 7/8, and 4/5 in order from least to greatest.

A) 4/5, 5/6, 7/8

B) 7/8, 5/6, 4/5

C) 5/6, 4/5, 7/8

D) They're all in a fraction show.

10. Arrange the fractions 7/10, 9/10, and 8/10 from least to greatest.

A) 7/10, 8/10, 9/10

B) 9/10, 7/10, 8/10

C) 8/10, 7/10, 9/10

D) Cannot be determined.

Addition or Subtraction of Fractions

11. Calculate the sum of 1/4 + 3/8.

A) 5/8

B) 7/8

C) 1/2

D) Cannot be determined.

12. Solve for 2/3 + 1/6.

A) 5/6

B) 1/2

C) 3/4

D) 2/3

13. What is 1/5 + 1/10?

A) 3/10

B) 2/5

C) 1/4

D) 1/5

14. Subtract 5/6 - 1/3.

A) 1/2

B) 1/3

C) 1/6

D) 5/6

15. Calculate 8 - 2 1/4.

A) 5 3/4

B) 6

C) 5 1/4

D) 8

16. Solve for 3 1/2 - 1 3/4.

A) 1 3/4

B) 1 4/5

C) 1 1/2

D) 3 1/2 is just trying to find a balance in life.

Practical Fraction Problems

17. Luis wants to determine the distance around the field. He knows that the width measures 140 3/8 ft and the length measures 380 1/4 ft. What is the perimeter of the field?

A) 1041 1/4 ft

B) 1040 1/2 ft

C) 1040 5/8 ft

D) 1041 3/4 ft

18. Marc wants to find out how much pizza is left. He had ordered 20 pizzas, and his friends ate 7 5/8 pizzas. How much pizza is left over?

A) 12 2/8 pizzas

B) 12 3/8 pizzas

C) 11 3/8 pizzas

D) 10 1/4 pizzas

19. Aaron worked 10 3/4 hours one day and 8 2/3 hours the next day. How many hours has he worked so far this week?

A) 18 5/12 hours

B) 19 5/12 hours

C) 19 1/6 hours

D) Aaron is sleeping at work.

20. If Aaron works 40 hours a week, how many more hours does he need to work to complete his work week?

A) 20 7/12 hours

B) 21 1/4 hours

C) 22 1/3 hours

D) Aaron is just trying to avoid the boss.

Unit 5: Unit Plan Making Sense of Fraction Multiplication and Division

5.NR.3: Describe fractions and perform operations with fractions to solve real-life mathematical problems using part-whole strategies and visual models.

Grade: 5th	Subject: Math

Domain: Fractions	Big Idea: Numerical Reasoning	Standard Topic: Fractions
5.NR.3.1 Explain the meaning of a fraction as division of the numerator by the denominator (a/b= a ÷ b). Solve problems involving division of whole numbers leading to answers in the form of fractions or mixed numbers. **5.NR.3.4 Model and solve problems involving multiplication of a fraction and a whole number.** **5.NR.3.5 Explain why multiplying a whole number by a fraction greater than one results in a product greater than the whole number, and why multiplying a whole number by a fraction less than one results in a product less than the whole number and multiplying a whole number by a fraction equal to one results in a product equal to the whole number.** **5.NR.3.6 Model and solve problems involving division of a unit fraction by a whole number and a whole number by a unit fraction.**		

Know Now: What students should know before class.	Do Now: What students should be able to do.
Students should know that a fraction is a quotient. Should be able to multiply a whole number by a fraction.	Students should be able to describe fractions and solve fraction problems by performing operations with fractions. Students should be able to solve problems with fractions greater or less than a whole number. Create unit fractions. Ask questions based on gathered information, observations, and displays to solve relevant problems.

Learning Targets
Create a sense of computational fluency. Students should be able to make sense of fractions by using visuals. They should be able to add fractions with unlike denominators. Generate equivalent fractions. Students should be able to investigate lengths and measurements on the number line using fractions. Ask statistical questions involving fractions.

Instructional Strategies
Compare fractions as subsets of a whole. Be able to create fractional parts. Order fractions on a number line. Make sense of division with fractions with fractional divisors and dividends. Review the concept of estimation. Have students consider if a sum or a difference of fractions are reasonable. Ask and answer questions involving measurement and graphical displays of data. Explore real-world applications.

Evidence of Learning
Students are able to see fractions as parts of a whole. Create their own set of fractions. Multiply and divide fractions that other students create. Find the unit fractions in real-world applications. Adding and subtracting fractions with like and unlike denominators.

Unit 5 5.NR.3.1 Lesson Plan

5.NR.3: Describe fractions and perform operations with fractions to solve real-life mathematical problems using part-whole strategies and visual models.

Grade: 5th	Subject: Math

Domain: Fractions	Big Idea: Numerical Reasoning	Standard Topic: Fractions
5.NR.3.1 Explain the meaning of a fraction as division of the numerator by the denominator (a/b = a ÷ b). Solve problems involving division of whole numbers leading to answers in the form of fractions or mixed numbers.		

Know Now: What students should know before class.	Do Now: What students should be able to do.
Students should be able to draw on prior knowledge that a fraction is a quotient. Division of the numerator and denominator. Use multiplication of fractions to create equal amounts.	Students should be able to apply and extend what they learn by using multiplication and division. They will multiply a whole number by a fraction or mixed number using properties of operations. Division of a whole number and a unit fraction.

Learning Targets
Students will multiply and divide fractions to solve practical problems. Solve problems involving division of whole numbers leading to answers in the form of fractions or mixed numbers. Students will gather information and data to be able to solve customary units of measure problems.

Instructional Strategies
Objective: **Warm-up: In two sentences, explain what you know about dividing a numerator by the denominator. Give an example.** **Jada was working on her homework. She asked her brother to help her divide 5 into 24. He said to make the 24 as the divisor and the 5 as the dividend. He explained that the answer would be a fraction with a remainder since 5 doesn't go evenly into 24.** **Is he correct? Justify your answer.** **Review: Parts of division and the concept of a remainder.** **Activity: Use the Unit 3 PowerPoint for the assignment 5.NR.3.1 is helpful for students to begin to understand dividing fractions.**

Assessment: Tell your neighbor how to divide a numerator by a denominator.

Unit 5: NR.3.4 Lesson Plan Model and solve problems involving multiplication of a fraction and a whole number.

5.NR.3: Describe fractions and perform operations with fractions to solve real-life mathematical problems using part-whole strategies and visual models.

Grade: 5th	Subject: Math

Domain: Fractions	Big Idea: Numerical Reasoning	Standard Topic: Fractions
5.NR.3.4 Model and solve problems involving multiplication of a fraction and a whole number.		

Know Now: What students should know before class.	Do Now: What students should be able to do.
Students should be able to draw on prior knowledge that a fraction is a quotient. Division of the numerator and denominator. Use multiplication of fractions to create equal amounts.	Students should be able to apply and extend what they learn by using multiplication and division. They will multiply a whole number by a fraction or mixed number using properties of operations. Model problems involving multiplication of a fraction and a whole number. Division of a whole number and a unit fraction.

Learning Targets
Students will multiply and divide fractions to solve practical problems. Solve problems involving multiplication of whole numbers leading to answers in the form of fractions or mixed numbers. Students will gather information and data to be able to solve customary units of measure problems.

Instructional Strategies
Objective: **Warm-up: In two sentences, explain what you know about multiplying a fraction by a whole number.** **Jada was working on her homework. She asked her brother to help her divide 5 into 24. He said to make the 24 as the divisor and the 5 as the dividend. He explained that the answer would be a fraction with a remainder since 5 doesn't go evenly into 24.** **Is he correct? Justify your answer.** **Review: Parts of division and the concept of a remainder.**

Activity: Use the Unit 3 PowerPoint for the assignment 5.NR.3.4 is helpful for students to begin to understand dividing fractions.
Assessment: Tell your neighbor how to divide a numerator by a denominator and how to multiply a whole number by a fraction.

Unit 5: NR.3.5 Lesson Plan Explain why multiplying a whole number by a fraction greater than one results in a product greater than the whole number, and why multiplying a whole number by a fraction less than one results in a product less than the whole number and multiplying a whole number by a fraction equal to one results in a product equal to the whole number.

5.NR.3: Describe fractions and perform operations with fractions to solve real-life mathematical problems using part-whole strategies and visual models.

Grade: 5th	Subject: Math

Domain: Fractions	Big Idea: Numerical Reasoning	Standard Topic: Fractions
5.NR.3.5 Explain why multiplying a whole number by a fraction greater than one results in a product greater than the whole number, and why multiplying a whole number by a fraction less than one results in a product less than the whole number and multiplying a whole number by a fraction equal to one result in a product equal to the whole number.		

Know Now: What students should know before class.	**Do Now: What students should be able to do.**
Students should be able to draw on prior knowledge that a fraction is a quotient. Division of the numerator and denominator. Use multiplication of fractions to create equal amounts.	**Students should be able to apply and extend what they learn by using multiplication and division. They will multiply a whole number by a fraction less than 1 and the same number mixed number greater than 1 using properties of operations. Model problems involving multiplication of a fraction and a whole number.**

Learning Targets
Students will multiply and divide fractions by a number less than 1, and multiply the same fraction by a number greater than 1 to solve problems. Solve problems involving multiplication of whole numbers leading to answers in the form of fractions or mixed numbers.

Instructional Strategies
Objective: **Warm-up: In two sentences, explain what you know about multiplying a fraction by a number less than 1, and multiplying the same fraction by a number greater than 1.** **Jada was working on her homework. She asked her brother to help her divide 5 into 24. He said to make the 24 as the divisor and the 5 as the dividend. He explained that the answer would be a fraction with a remainder since 5 doesn't go evenly into 24.** **Is he correct? Justify your answer.** **Review: Parts of division and the concept of a remainder.**

Activity: Use the Unit 3 PowerPoint for the assignment 5.NR.3.4 is helpful for students to begin to understand dividing fractions.
Assessment: Tell your neighbor how to divide a numerator by a denominator.

Objective: **Warm-up: In two sentences, explain what you know about dividing a unit fraction by a whole number and a whole number by a unit fraction.**	**Technology: PowerPoint is an option** **Materials: Paper, pencil.**

Unit 5 5.NR.3.6 Lesson Plan:

5.NR.3: Describe fractions and perform operations with fractions to solve real-life mathematical problems using part-whole strategies and visual models.

Day 1-2

5.NR.3.6 Model and solve problems involving division of a unit fraction by a whole number and a whole number by a unit fraction.	**Activity: Review dividing and multiplying (Dok 1)**

Jax was working at his new job. He made $400 for working 40 hours that week. What is the unit fraction of his pay? How do you know? **Review: Parts of division and the concept of a remainder.** **Activity: Use the Unit 3 PowerPoint for the assignment 5.NR.3.6 is helpful for students to begin to understand dividing fractions.**	
Assessment: Tell your neighbor how to find a unit fraction.	

Unit 5 Test

Understanding Fractions

1. A denominator divided by a numerator is a fraction.

A) True

B) False

2. Joan has 36 apple pies that she wants to divide between her and her five friends. How many apple pies can each have if each receives the same amount?

A) 6

B) 5

C) 7

D) 8

3. Kaynan has a Burrito box. He bought 30 tacos for $18. How much did each taco cost?

A) $0.50

B) $0.60

C) $0.75

D) $0.80

Division Problems

4. What is 72 divided by 12?

A) 5

B) 6

C) 7

D) 8

5. What is 80 divided by 5?

A) 14

B) 15

C) 16

D) 17

6. What is 145 divided by 6?

A) 23

B) 24

C) 25

D) 26

7. What is 360 divided by 18?

A) 18

B) 19

C) 20

D) 21

8. What is 170 divided by 20?

A) 8

B) 8 1/2

C) 9

D) 9 1/2

9. What is 185 divided by 15?

A) 12

B) 12 1/2

C) 13

D) 13 1/2

Multiplying Whole Numbers by Fractions

10. 8 × (1/2) = ______

A) 3

B) 4

C) 5

D) 6

11. 48 × (1/4) = ______

A) 10

B) 11

C) 12

D) 13

12. 36 × (3/4) = ______

A) 24

B) 25

C) 26

D) 27

13. 18 × (5/6) = ______

A) 12

B) 13

C) 14

D) 15

14. 40 × (1/5) = ______

A) 6

B) 7

C) 8

D) 9

15. 100 × (3/5) = ______

A) 50

B) 55

C) 60

D) 65

Conclusion:

16. Conclusion: When a whole number is multiplied by a fraction less than 1, the result is

A) greater

B) smaller

17. When a whole number is multiplied by a fraction greater than 1, the result is

A) greater

B) smaller

Unit Fractions

18. There are 60 minutes in an hour. The unit fraction of a minute is

A) 1/30

B) 1/45

C) 1/60

D) 1/90

19. Jason works at a local store. He brought home $420 a week. What is the unit fraction for hours worked at the store if he works 40 hours a week?

A) 1/10

B) 1/15

C) 1/20

D) 1/25

20. Conner fills up his gas tank, where the unit fraction is $2.80. If he buys 20 units, how much does it cost him?

A) $50

B) $56

C) $60

D) $70

21. Travis bought a dozen cookies for $2.40. What is the unit fraction for a cookie?

A) $0.10

B) $0.15

C) $0.20

D) $0.25

22. Maria wants to take her friends to a game. If the unit fraction for tickets at the Falcons game is 1/$34, how much would five tickets cost?

A) $170

B) $160

C) $150

D) $140

Unit 6: Unit Plan Extending Place Value and Working with Decimals to Solve Problems

Unit 6 Standard: Extending Place Vale and Working with Decimals to Solve Problems

5.NR.4: Read, write, and compare decimal numbers to the thousandths place, and round and perform operations with decimal numbers to the hundredths place to solve real-life mathematical problems.

5.MDR.7: Solve problems involving customary measurements, metric measurements, and time and analyze graphical displays of data to answer relevant questions.

Grade: 5th	Subject: Math

Domain: Decimals	Big Idea: Numerical Reasoning	Standard Topic: Decimals
5.NR.4.1 Read and write decimal numbers to the thousandths place using base-ten numerals written in standard form and expanded form. **5.NR.4.2 Represent, compare, and order decimal numbers to the thousandths place based on the meanings of the digits in each place, using >, =, and < symbols to record the results of comparisons.** **5.NR.4.3 Use place value understanding to round decimal numbers to the hundredths place.** **5.NR.4.4 Solve problems involving addition and subtraction of decimal numbers to the hundredths place using a variety of strategies.** **5.MDR.7.2 Ask questions and answer them based on gathered information, observations, and appropriate graphical displays to solve problems relevant to everyday life.**		

Know Now: What students should know before class.	Do Now: What students should be able to do.
Students should know about numerical operations with whole numbers.	Students should be able to use their understanding of place value to compare, locate, and order decimals to the tenths and hundredths place. Students will use their knowledge of place value to add and subtract decimals.

Learning Targets
Create a sense of computational fluency. Students should be able to make sense of decimals by using visuals. They should be able to compare, locate, and order decimals. Students should be able to investigate lengths and measurements using decimals. Ask statistical questions involving decimals.

Instructional Strategies
Read and write decimals to the thousandths place using base ten numbers written in expanded and standard form. **Compare decimals to the thousandths place.** **Order decimals on the number line.** **Round decimals to the hundredths place by using place value understanding.** **Solve problems by adding and subtracting decimals to the hundredths place.** **Ask and answer questions involving measurement and graphical displays of data. Explore real-world applications.**

Evidence of Learning
Students are able to see decimals as parts of a whole. Create their own stories and scenarios using decimals. Add and subtract decimals that other students create.

Unit 6: Unit 6 5.NR.4.1 Lesson Plan:

5.NR.4: Read, write, and compare decimal numbers to the thousandths place, and round and perform operations with decimal numbers to the hundredths place to solve real-life mathematical problems.

5.MDR.7: Solve problems involving customary measurements, metric measurements, and time and analyze graphical displays of data to answer relevant questions.

Grade: 5th	Subject: Math

Domain: Decimals	Big Idea: Numerical Reasoning	Standard Topic: Decimals
5.NR.4.1 Read and write decimal numbers to the thousandths place using base-ten numerals written in standard form and expanded form.		

Know Now: What students should know before class.	Do Now: What students should be able to do.
Students should know about numerical operations with whole numbers.	Students should be able to apply and extend what they learn by using reading and writing decimals.

Learning Targets
Reading and writing decimals to the thousandths place.

Instructional Strategies
Objective: **Warm-up: In two sentences, explain what you remember in previous discussions about decimals.** **Hector is trying out for the swim team. He wants to beat the record set by a teammate of 50m in 30.345 mins. If the first reasonable goal is to beat the whole number of mins. How many mins would that be? Round to the nearest tenth?** **Does anyone remember from a previous class or experience how to round decimals? If yes, explain and give a couple of mins of background knowledge.**
Review: Rounding decimals to the whole, tenth, hundredth, and thousandth place. Activity: Use the Unit 6 PowerPoint to The assignment 5.NR.4.1 is helpful for students to begin to understand reading and writing decimals in expanded and standard form.
Assessment: Write two decimal numbers. Pass them to your neighbor. See if you can round the decimals to the whole, tenth, hundredth, and thousandth place.

Unit 6: Unit 6 5.NR.4.2 Lesson Plan:

6.NR.4: Read, write, and compare decimal numbers to the thousandths place, and round and perform operations with decimal numbers to the hundredths place to solve real-life mathematical problems.

5.MDR.7: Solve problems involving customary measurements, metric measurements, and time and analyze graphical displays of data to answer relevant questions.

Grade: 5th	Subject: Math

Domain: Decimals	Big Idea: Numerical Reasoning	Standard Topic: Decimals
5.NR.4.2 Represent, compare, and order decimal numbers to the thousandths place based on the meanings of the digits in each place, using >, =, and < symbols to record the results of comparisons.		

Know Now: What students should know before class.	Do Now: What students should be able to do.
Students should know about numerical comparisons with whole numbers.	Students should be able to apply and extend what they learn by representing, comparing, and ordering decimals.

Learning Targets
Represent, compare, and order decimals to the thousandths place.

Instructional Strategies
Objective: **Warm-up: In two sentences, explain what you remember about rounding decimals.** **Victor is trying out for the archery team. He wants to be competitive. He is considering hitting targets from 50.5, 50.55, and 49.009 m away. List the distance from the targets from least to greatest.** **Does anyone remember from a previous class or experience about ordering decimals? If yes, explain and give a couple of mins of background knowledge.**

Review: Rounding decimals to the whole, tenth, hundredth, and thousandth place. Add ordering to the review. **Activity: Use the Unit 6 PowerPoint to The assignment 5.NR.4.2 is helpful for students to begin to understand comparing and ordering decimals.**
Assessment: Tell your neighbor how to order three decimals. And/or ask your neighbor to write down three decimals, swap papers, and order the decimals that your neighbor has written down.

Unit 6: Unit 6 5.NR.4.3 Lesson Plan:

7.NR.4: Read, write, and compare decimal numbers to the thousandths place, and round and perform operations with decimal numbers to the hundredths place to solve real-life mathematical problems.

5.MDR.7: Solve problems involving customary measurements, metric measurements, and time and analyze graphical displays of data to answer relevant questions.

Grade: 5th	Subject: Math

Domain: Decimals	Big Idea: Numerical Reasoning	Standard Topic: Decimals
5.NR.4.3 Use place value understanding to round decimal numbers to the hundredths place.		

Know Now: What students should know before class.	Do Now: What students should be able to do.
Students should know about numerical comparisons with whole numbers.	Students should be able to apply and extend what they learn by rounding decimals.

Learning Targets
Round decimals to the hundredths place.

Instructional Strategies
Objective: **Warm-up: In two sentences, explain what you remember about rounding decimals.** **Victor is trying out for the archery team. He wants to be competitive. He is considering hitting targets from 50.5, 50.55, and 49.009 m away. List the distance from the targets from least to greatest.** **Does anyone remember from a previous class or experience about ordering decimals? If yes, explain and give a couple of mins of background knowledge.**
Review: Rounding decimals to the whole, tenth, hundredth, and thousandth place. Add ordering to the review. **Activity: Use the Unit 6 PowerPoint to The assignment 5.NR.4.2 is helpful for students to begin to round decimals.**
Assessment: Tell your neighbor how to order three decimals. And/or ask your neighbor to write down three decimals, swap papers, and round the decimals that your neighbor has written down to the nearest whole, tenth, and hundredth.

Unit 6: Unit 6 5.NR.4.3 Lesson Plan:

7.NR.4: Read, write, and compare decimal numbers to the thousandths place, and round and perform operations with decimal numbers to the hundredths place to solve real-life mathematical problems.

5.MDR.7: Solve problems involving customary measurements, metric measurements, and time and analyze graphical displays of data to answer relevant questions.

Grade: 5th	Subject: Math

Domain: Decimals	Big Idea: Numerical Reasoning	Standard Topic: Decimals
5.NR.4.3 Use place value understanding to round decimal numbers to the hundredths place.		

Know Now: What students should know before class.	Do Now: What students should be able to do.
Students should know about numerical comparisons with whole numbers.	**Students should be able to apply and extend what they learn by rounding decimals.**

Learning Targets
Round decimals to the hundredths place.

Instructional Strategies
Objective: **Warm-up: In two sentences, explain what you remember about rounding decimals.** **Victor is trying out for the archery team. He wants to be competitive. He is considering hitting targets from 50.5, 50.55, and 49.009 m away. List the distance from the targets from least to greatest.** **Does anyone remember from a previous class or experience about ordering decimals? If yes, explain and give a couple of mins of background knowledge.**
Review: Rounding decimals to the whole, tenth, hundredth, and thousandth place. Add ordering to the review. **Activity: Use the Unit 6 PowerPoint to The assignment 5.NR.4.2 is helpful for students to begin to round decimals.**
Assessment: Tell your neighbor how to order three decimals. And/or ask your neighbor to write down three decimals, swap papers, and round the decimals that your neighbor has written down to the nearest whole, tenth, and hundredth.

Unit 6: Unit 6 5.NR.4.4 Lesson Plan:

8.NR.4: Read, write, and compare decimal numbers to the thousandths place, and round and perform operations with decimal numbers to the hundredths place to solve real-life mathematical problems.

5.MDR.7: Solve problems involving customary measurements, metric measurements, and time and analyze graphical displays of data to answer relevant questions.

Grade: 5th	Subject: Math

Domain: Decimals	Big Idea: Numerical Reasoning	Standard Topic: Decimals
5.NR.4.4 Solve problems involving addition and subtraction of decimal numbers to the hundredths place using a variety of strategies.		

Know Now: What students should know before class.	Do Now: What students should be able to do.
Students should know about adding and subtracting with whole numbers.	Students should be able to apply and extend what they learn by adding and subtracting decimals.

Learning Targets
Adding and subtracting decimals.

Instructional Strategies
Objective: **Warm-up: In two sentences, explain what you remember about adding and subtracting decimals.** **Tracy wants to add two decimals together. What is one idea that you could give him?** **Does anyone remember from a previous class or experience about adding decimals? If yes, explain and give a couple of mins of background knowledge.**
Review: Rounding decimals to the whole, tenth, hundredth, and thousandth place. Add adding and subtracting decimals to the review. **Activity: Use the Unit 6 PowerPoint to The assignment 5.NR.4.4 is helpful for students to begin to understand adding and subtracting decimals.**
Assessment: Tell your neighbor how to write down two decimals. And/or ask your neighbor to write down two decimals and swap papers. Add and subtract the decimals that your neighbor has written down. Check your work by completing the opposite direction.

Unit 5 Test

Explain the meaning of a fraction. Solve problems involving division of whole numbers.

1. A denominator divided by a numerator is a fraction. True or False?

2. Joan has 36 apple pies that she wants to divide between her and her five friends. How many apple pies can each have if each receive the same amount?

3. Kaynan has a Burrito box. He bought 30 tacos for $18. How much did each taco cost? Or what is the unit fraction?

Divide.

4. 72/12

5. 80/5

6. 145/6

7. 360/18

8. 170/20

9. 185/15

Multiplying a whole number by a fraction. Tell whether the whole number is bigger than the fraction, results in the number being greater than the number and whether the results are smaller if the number is multiplied by a fraction less than 1.

10. 8× = ______

11. 48× = ______

12. 36× = ______

13. 18× = _____

14. 40× = _____

15. 100× = _____

Conclusion: When a whole number is multiplied by a fraction less than 1, the fraction is ?_______ When a whole number is multiplied by a fraction greater than 1, the fraction is ?_______

16. There are 60 minutes in an hour. The unit fraction of a minute is _________.

17. Jason works at a local store. He brought home $420 a week. What is the unit fraction for hours worked at the store?

18. Conner fills up his gas tank, where the unit fraction is $2.80. If he buys 20 units How much does it cost him?

19. Travis bought a dozen cookies for $2.40. What is the unit fraction for a cookie?

20. Maria wants to take her friends to a game. If the unit fraction for a ticket at the Falcons game is 1/$34, how much would five tickets cost?

Unit 6: Unit 6 5.NR.4.4 Lesson Plan:

9.NR.4: Read, write, and compare decimal numbers to the thousandths place, and round and perform operations with decimal numbers to the hundredths place to solve real-life mathematical problems.

5.MDR.7: Solve problems involving customary measurements, metric measurements, and time and analyze graphical displays of data to answer relevant questions.

Grade: 5th	Subject: Math

Domain: Decimals	Big Idea: Numerical Reasoning	Standard Topic: Decimals
5.MDR.7.2 Ask questions and answer them based on gathered information, observations, and appropriate graphical displays to solve problems relevant to everyday life.		

Know Now: What students should know before class.	Do Now: What students should be able to do.
Students should know about reading, comparing, and ordering decimals. Students should know about adding and subtracting with whole numbers.	Students should be able to add and subtract decimals in real-world situations. Students should be able to apply and extend what they learn by adding and subtracting decimals.

Learning Targets
Reading, comparing, and ordering decimals. Adding and subtracting decimals.

Instructional Strategies
Objective: Warm-up: In two sentences, explain what you remember about adding and subtracting decimals. Sean is considering buying a meal from ZipIn. He wants to buy a sandwich and a cola. The sign says that the sandwich is \$2.25 and the cola is \$ 3.75. Does he have enough money if he takes a \$5 bill from his wallet? Does anyone remember anything, from a different class or experience, about adding and subtracting decimals? If yes, explain and give a couple of mins of background knowledge.

Review: Rounding, ordering, adding, and subtracting decimals. Activity: Use the Unit 6 PowerPoint for the assignment 5.MDR.7.2 Word problems are helpful for students to begin to understand adding and subtracting decimals in real-world situations.
Assessment: Tell your neighbor how to write down two decimals. And/or ask neighbor to write down two decimals and swap papers. Add and subtract the decimals that your neighbor has written down. Check your work by completing the opposite direction.

Unit 6 Test: Decimals

Read and Write Decimal Numbers

1. How would you write "Four thousand eight hundred sixty-two and fifty-three hundredths" as a decimal?

A. 4,862.053

B. 4,862.53

C. 4,862.503

D. 48,620.53

2. How would you write "Two hundred ninety-three and eighty-six hundredths" as a decimal?

A. 293.086

B. 293.86

C. 2,938.6

D. 29.386

3. How would you write "Ten thousand three hundred seventy-five and six hundred twenty-eight thousandths" as a decimal?

A. 10,375.0628

B. 10,375.628

C. 10,375.062

D. 1,037.5628

4. How would you write "Eight thousand five hundred ninety-seven and three hundred sixty-four thousandths" as a decimal?

A. 8,597.364

B. 8,597.0364

C. 85,973.64

D. 8,597.00364

Change Decimals into Words

5. How would you read the decimal 5863.41 in words?

A. Five thousand eight hundred sixty-three and forty-one hundredths

B. Fifty-eight six hundred thirty-four and ten

C. Five hundred eighty-six and three hundred forty-one thousandths

D. Fifty-eight thousand six hundred thirty-four and ten

6. How would you read the decimal 381.352 in words?

A. Three hundred eighty-one and three hundred fifty-two thousandths

B. Thirty-eight and one hundred three hundred fifty-two

C. Three hundred eighty-one and fifty-three hundredths

D. Three hundred and eighty-one point three five two

7. How would you read the decimal 0.572 in words?

A. Zero point five seventy-two

B. Zero and five hundred seventy-two thousandths

C. Five hundred seventy-two

D. Fifty-seven and two tenths

8. How would you read the decimal 19,234.75 in words?

A. One thousand nine hundred twenty-three and four hundred seventy-five

B. Nineteen thousand two hundred thirty-four and seventy-five hundredths

C. Nineteen and two hundred thirty-four point seven five

D. Nineteen thousand two hundred and thirty-four and seventy-five

Compare Decimals

9. Compare: 483.23 ____ 485.92

A. >

B. <

C. =

10. Compare: 49,093.38 ____ 9,384.332

A. >

B. <

C. =

11. Compare: 848.3 ____ 892.32

A. >

B. <

C. =

12. Compare: 6554.78 ____ 673.92

A. >

B. <

C. =

Order Decimals from Least to Greatest

13. Order: 381.45, 3,891.32, 392.34

A. 381.45, 392.34, 3,891.32

B. 392.34, 381.45, 3,891.32

C. 3,891.32, 381.45, 392.34

D. 381.45, 3,891.32, 392.34

14. Order: 582.34, 38.352, 38,349.21

A. 38.352, 582.34, 38,349.21

B. 582.34, 38.352, 38,349.21

C. 38,349.21, 582.34, 38.352

D. 38,349.21, 38.352, 582.34

15. Order: 83,421.2, 382.212, 29,352.12

A. 382.212, 29,352.12, 83,421.2

B. 83,421.2, 382.212, 29,352.12

C. 29,352.12, 382.212, 83,421.2

D. 382.212, 83,421.2, 29,352.12

16. Order: 7,436.23, 3,201.23, 382.32

A. 382.32, 3,201.23, 7,436.23

B. 7,436.23, 3,201.23, 382.32

C. 3,201.23, 382.32, 7,436.23

D. 382.32, 7,436.23, 3,201.23

Add the Decimals

17. What is 546.72 + 3949.31?

A. 4,496.03

B. 4,495.93

C. 4,495.03

D. 4,500.03

18. What is 5859.33 + 7362.47?

A. 13,221.80

B. 13,221.90

C. 13,221.70

D. 13,230.80

19. What is 423.9 + 39.087?

A. 462.987

B. 463.987

C. 462.977

D. 462.097

20. What is 9,850.52 + 437.602?

A. 10,288.122

B. 10,288.12

C. 10,287.122

D. 10,287.12

Subtract the Decimals

21. What is 827.32 - 392.39?

A. 434.93

B. 434.83

C. 434.43

D. 434.53

22. What is 7322.98 - 5837.77?

A. 1,485.21

B. 1,485.11

C. 1,485.31

D. 1,485.41

23. What is 38292.87 - 3829.56?

A. 34,463.31

B. 34,463.41

C. 34,463.21

D. 34,463.11

24. What is 5788.54 - 2765.54?

A. 3,023.00

B. 3,022.00

C. 3,023.10

D. 3,022.10

Word Problem

25. Eric wants to go to the mall to buy an outfit for the dance. He has $300. He buys a shirt for $79.50 and a pair of pants for $85.76. He finds a pair of shoes for $149.53. Does he have enough money to buy his outfit?

A. Yes, he will have $5.21 left.

B. No, he will be $14.79 short.

C. Yes, he will have $14.79 left.

D. No, he will be $5.21 short.

Unit 7 Standard: Exploring Geometry and the Coordinate Plane

5.PAR.6: Solve real-life problems by creating and analyzing numerical patterns using the given rule(s).

5.GSR.8: Examine properties of polygons and rectangular prisms, classify polygons by their properties, and discover volume of right rectangular prisms.

Grade: 5th	**Subject: Math**

Domain: Coordinate Plane	**Big Idea:** Patterning & Algebraic Reasoning and Geometric & Spatial Reasoning	**Standard Topic: Coordinate Plane**
5.PAR.6.1 Generate two numerical patterns using two given rules. Identify apparent relationships between corresponding terms by completing a table. **5.PAR.6.2 Represent problems by plotting ordered pairs and explain coordinate values of points in the first quadrant of the coordinate plane.** **5.GSR.8.1 Classify, compare, and contrast polygons based on properties.** **5.GSR.8.2 Determine, through exploration and investigation, that attributes belonging to a category of two-dimensional figures also belong to all subcategories of that category.**		

Know Now: What students should know before class.	Do Now: What students should be able to do.
Students should know about numerical operations with whole numbers. Students should know how to recognize some polygons.	Students will build on their knowledge of recognizing polygons. Students should know how to recognize patterns in math. They should know how to identify relationships between corresponding terms.

Learning Targets
Recognize patterns in math and shapes. Students should be able to recognize which Quadrant points are located in. Students should be able to plot points on the coordinate plane. Ask questions about polygons.

Instructional Strategies
Recognize mathematical patterns and identify relationships of the patterns. Consider points plotted in the coordinate plane and identify the ordered pairs. Determine the quadrants that points are located in. Classify, compare, and contrast polygons. Ask and answer questions involving measurement and graphical displays of data. Explore real-world applications.

Evidence of Learning
Students are able to plot points of their own. Create their own stories and scenarios using the coordinate plane. Locate points that other students have plotted on a coordinate plane by using given ordered pairs.

Unit 7: 5.PAR.6.1 Lesson Plan:

5.PAR.6: Solve real-life problems by creating and analyzing numerical patterns using the given rule(s).

5.GSR.8: Examine properties of polygons and rectangular prisms, classify polygons by their properties, and discover volume of right rectangular prisms.

Grade: 5th	Subject: Math

Domain: Geometry & the Coordinate Plane	Big Idea: Patterning & Algebraic Reasoning and Geometric & Spatial Reasoning	Standard Topic: Analyzing numerical patterns.
5.PAR.6.1 Generate two numerical patterns using two given rules. Identify apparent relationships between corresponding terms by completing a table.		

Know Now: What students should know before class.	Do Now: What students should be able to do.
Students should have a general understanding of a numerical pattern.	**Students should be able to study patterns and come up with a rule. Given a rule, list the numbers generated by applying the rule.**

Learning Targets
Look at two number patterns. Check for relationships within the two patterns.

Instructional Strategies
Objective: **Warm-up: In two sentences, explain what you remember about patterns and number patterns. Give an example.** **Jordan is working on his homework. He writes down the patterns: Pattern 1: 2, 4, 6, 8, 10 and Pattern 2: 3, 6, 9, 12, 15. His sister tells him that pattern 2 increases faster than pattern 1. Is she correct? Explain.**
Assessment: Tell your neighbor how to write down two patterns. And/or ask neighbor to write down two patterns and swap papers. Create a rule for the two patterns you have written. Check your work by completing the opposite direction.

Unit 7: Unit 7 5.PAR.6.2 Lesson Plan:

6.PAR.6: Solve real-life problems by creating and analyzing numerical patterns using the given rule(s).

5.GSR.8: Examine properties of polygons and rectangular prisms, classify polygons by their properties, and discover volume of right rectangular prisms.

Grade: 5th	Subject: Math

Domain: Geometry & the Coordinate Plane	Big Idea: Patterning & Algebraic Reasoning and Geometric & Spatial Reasoning	Standard Topic: Analyzing numerical patterns.
5.PAR.6.2 Represent problems by plotting ordered pairs and explain coordinate values of points in the first quadrant of the coordinate plane.		

Know Now: What students should know before class.	Do Now: What students should be able to do.
Students should have a general understanding of a numerical pair.	**Students should be able to plot ordered pairs and explain coordinate values of points in the first quadrant of the coordinate plane.**

Learning Targets
Look at two numbers in an ordered pair. Plot the ordered pair in the first quadrant of the coordinate plane. Given a point, state the ordered pair of that point.

Instructional Strategies
Objective: **Warm-up:** In two sentences, explain what you remember about ordered pairs and the coordinate plane. Taylor wants to play a game of battleship with you. He plots his ship on points (3,1), (3,2) and (3,3). You plot your ship on points (4,5), (5,5) and (6,5). Taylor says (6,8). Is that a hit or a miss? Explain. Review: Maps are a good way to explain plotting points in the coordinate plane. Often, remind the students that we are using just the 1st Quadrant to keep things easier than using the whole grid. (Consider a 10 by 10 or a 20 by 20 grid.)

Activity: Use the Unit 6 PowerPoint for the assignment. 5.MDR.7.2 is helpful for students to begin to understand the coordinate plane.
Assessment: Tell your neighbor to write down two ordered pairs. Plot the points on the coordinate plane. Check your work by completing the opposite direction.

Unit 7: 5.GSR.8.1 Lesson Plan:

7.PAR.6: Solve real-life problems by creating and analyzing numerical patterns using the given rule(s).

5.GSR.8: Examine properties of polygons and rectangular prisms, classify polygons by their properties, and discover volume of right rectangular prisms.

Grade: 5th	Subject: Math

Domain: Geometry & the Coordinate Plane	Big Idea: Patterning & Algebraic Reasoning and Geometric & Spatial Reasoning	Standard Topic: Analyzing numerical patterns.
5.GSR.8.1 Classify, compare, and contrast polygons based on properties.		

Know Now: What students should know before class.	Do Now: What students should be able to do.
Students should have a general understanding of a polygon.	**Students should be able to recognize a certain polygon. They should be able to classify, compare, and contrast polygons based on stated properties.**

Learning Targets
Given a polygon, classify, compare, and contrast the polygon according to its attributes.

Instructional Strategies
Objective: **Warm-up: In two sentences, explain what you remember about polygons.** **Are polygons tangible? Yes or No. Explain.** **Review: Properties of polygons. You want to ask AI to create an awesome creature with three polygons. Which ones would you choose and why? Example: a fox created using three triangles. You can create one by sketching, too.** **Activity: Use the Unit 6 PowerPoint for the assignment 5.MDR.7.2 is helpful for students to begin to understand the coordinate plane.**

Assessment: Tell your neighbor two polygons. Try to state one thing they have in common and one thing that makes them different.

Unit 7: 5.GSR.8.2 Lesson Plan:

8.PAR.6: Solve real-life problems by creating and analyzing numerical patterns using the given rule(s).

5.GSR.8: Examine properties of polygons and rectangular prisms, classify polygons by their properties, and discover volume of right rectangular prisms.

Grade: 5th	Subject: Math

Domain: Geometry & the Coordinate Plane	Big Idea: Patterning & Algebraic Reasoning and Geometric & Spatial Reasoning	Standard Topic: Comparing polygons
5.GSR.8.2 Determine, through exploration and investigation, that attributes belonging to a category of two-dimensional figures also belong to all subcategories of that category.		

Know Now: What students should know before class.	Do Now: What students should be able to do.
Students should have a general understanding of a polygon.	Students should be able to recognize a certain polygon. They should be able to classify, compare, and contrast polygons based on stated properties.

Learning Targets
Given a polygon, classify, compare, and contrast the polygon according to its attributes. Create subcategories of those polygons.

Instructional Strategies
Objective: **Warm-up: In two sentences, explain what you know about the attributes of polygons. Choose two:** **Ex. Circle: Has a radius. Has a diameter.** **Review: Attributes of given polygons, like triangles, have three sides. They have 180 degrees. Etc.** **Activity: Use the Unit 6 PowerPoint for the assignment 5.GSR.8.2 is helpful for students to begin to understand the attributes of polygons.**

Assessment: Tell your neighbor two polygons. Try to state one thing they have in common and one thing that makes them different. Try to come up with a subcategory also.

Unit 7 Test

1. Consider the following characteristics, which are valid for all polygons?

A. All sides have different lengths

B. All angles are congruent (equal).

C. Each side is a curve

D. The polygons are 3D

2. What is the sum of the interior angles of a triangle?

A. 100

B. 90

C. 95

D. 180

3. How many sides does an octagon have?

A. 4

B. 5

C. 7

D. 8

4. Which of the following have all equal angles and sides?

A. Scalane triangle

B. Square

C. Isosceles triangle

D. Irregular quadrilateral

5. Which quadrilateral has only one pair of parallel sides?

A. Triangle

B. Rhombus

C. Trapezoid

D. Rectangle

6. What is the next number of the pattern? 3, 6, 9, 12, 15

A. 21

B. 16

C. 17

D. 18

7. Identify the rule of the pattern. 625, 125, 25, 5

A. Multiply by 5

B. Divide by 5

C. Add 5

D. Subtract 5

8. What is the next shape in the following pattern? Star, rocket, star, rocket, rocket,

A. Star

B. Rocket

C. Rocket, rocket

D. Star, star

9. What is the name of the vertical axis in the coordinate plane?

A. Y-axis

B. X-axis

C. Origin

D. Rabbit

10. In which quadrant is the point (2,3) located?

A. 1st

B. 2nd

C. 3rd

D. 4th

11. If a point moves from (5,7) to (5,9), in which direction does it move?

A. Right

B. Left

C. Up

D. Down

12. Which of the following points are on the x-axis?

A. (0,7)

B. (5,2)

C. (4,0)

D. (3,3)

13. A line with no slope is called?

A. Horizontal

B. Vertical

C. Diagonal

D. Curved

14. If the pattern is "add 5", what is the 8th term starting at 1?

A. 17

B. 28

C. 43

D. 36

15. A shared attribute of rectangle and a rhombus is?

A. 4 sides

B. 3 sides

C. All equal sides

D. No equal sides

16-20 Graph the following points on a coordinate plane. (please add coordinate plane)

16. (6,5)

17. (8,2)

18. (4,3)

19. (8,8)

20. (7,1)

Post Test

Unit 8.

Expressions

Quiz 1

Simplify.

1. 4(10) + 3(6)

2. 3(8+2) + 4(3)

3. 5×3 - 4×1

4. (7×5)(4×3)

5. (13-3) ÷ 2

Quiz 2

Simplify.

1. 2(8) - 4^2

2. (3+3+3) - (2+2)

3. (18÷6) + (7+3)

4. (24+6)÷(8+2)

5. (9×8)÷(6×6)

Quiz 3

Simplify.

1. (53+7)(5+5)

2. (2×2×2)(2+3)

3. 6+4×1

4. 3(9)÷37(9)(0)

Unit 2 Comparing Numbers

Quiz 1

Compare. > , <, or =.

1. 8 7

2. 13 19

3. 61 59

4. 121 121

5. 86 68

Quiz 2

Compare > , <, or =.

1. 320 249

2. 745 801

3. 621 399

4. 243 362

5. 151 514

Quiz 3

Compare > , <, or =.

1. 278 781

2. 323 236

3. 142 424

4. 580 445

5. 121 315

Quiz 1

Write in standard form.

1. Two thousand one hundred twenty-five.

2. Nine hundred thirty-two.

3. Six thousand seven hundred forty-six.

4. Eighty thousand five hundred ninety-seven.

5. One thousand three hundred fifty-three.

Quiz 2

Write in standard form.

1. Three thousand four hundred seventy-two.

2. Four thousand three hundred fifty-six.

3. Seventy thousand six hundred and forty-seven.

4. Fifty thousand five hundred and fifty-five.

5. Three thousand eight hundred twenty-four.

Quiz 3

Write in standard form.

1. Two thousand three hundred twenty-five.

2. Four thousand eight hundred fifty-six.

3. One thousand one hundred eleven.

4. Three hundred forty-seven.

5. Six thousand one hundred eighty-five.

Multiplication: Three Digits by Two Digits

Quiz 1

Multiply

1. 389	**2.** 678	**3.** 144	**4.** 286	**5.** 965
× 24	× 12	× 20	× 81	× 37

Quiz 2

Multiply

1. 324	**2.** 963	**3.** 146	**4.** 574	**5.** 829
× 15	× 21	× 12	× 39	× 16

Quiz 3

Multiply

1. 563	**2.** 211	**3.** 349	**4.** 131	**5.** 359
× 17	× 34	× 47	× 25	× 28

Multiplication: Three Digits by Three Digits

Quiz 1

Multiply

1. 773	**2.** 614	**3.** 103	**4.** 234	**5.** 764
× 289	× 471	× 383	× 432	× 221

Quiz 2

Multiply

1. 250	**2.** 314	**3.** 177	**4.** 864	**5.** 757
× 159	× 500	× 211	× 252	× 164

Quiz 3

Multiply

1. 103	**2.** 232	**3.** 318	**4.** 142	**5.** 682
× 761	× 250	× 525	× 419	× 127

Standard Form

Quiz 1

Write the number in standard form.

1. 20,000 + 8,000 + 500 + 40 + 2

2. 30,000 + 7,000 + 600 + 50 + 3

3. 80,000 + 6,000 + 700 + 30 + 5

4. 5,000 + 300 + 80 + 1

5. 70,000 + 800 + 60 + 2

Quiz 2

Write the number in standard form.

1. 7,000 + 700 + 70 + 1

2. 30,000 + 9,000 + 800 + 40 + 5

3. 4,000 + 600 + 20 + 1

4. 50,000 + 600 + 80 + 2

5. 20,000 + 5,000 + 70 + 6

Quiz 3

Write the number in standard form.

1. 60,000 + 5,000 + 80 + 4

2. 70,000 + 60 + 5

3. 30,000 + 7,000 + 400 + 60 + 3

4. 40,000 + 8,000 + 50 + 2

5. 80,000 + 9,000 + 400 + 7

Multiplication: Fraction by Fraction

Quiz 1

Multiply.

1. $\frac{7}{8} \times \frac{4}{11}$ 2. $\frac{3}{5} \times \frac{20}{4}$ 3. $\frac{8}{14} \times \frac{7}{21}$ 4. $\frac{4}{5} \times \frac{2}{4}$ 5. $\frac{6}{32} \times \frac{8}{3}$

Quiz 2

Multiply.

1. $\frac{1}{2} \times \frac{1}{4}$ 2. $\frac{3}{5} \times \frac{20}{4}$ 3. $\frac{8}{14} \times \frac{7}{21}$ 4. $\frac{4}{5} \times \frac{2}{4}$ 5. $\frac{6}{32} \times \frac{8}{3}$

Quiz 3

Multiply.

1. $\frac{3}{9} \times \frac{3}{7}$ 2. $\frac{4}{5} \times \frac{80}{4}$ 3. $\frac{8}{14} \times \frac{7}{24}$ 4. $\frac{4}{32} \times \frac{6}{36}$ 5. $\frac{6}{27} \times \frac{9}{21}$

Adding Decimals

Quiz 1

1. 87.32+16.34
2. 127.34+234.63
3. 5872.43+8237.23
4. 843.81+4638.32
5. 34.389+376.34

Quiz 2

1. 0.98+1.34
2. 0.234+0.25
3. 837.35+339.21
4. 892.28+283.21
5. 2847.32+382.32

Quiz 3

1. 3984.34+9384.23
2. 3749.93+203.57
3. 2839.91+392.21
4. 2938.324+23.38
5. 1830.38+3849.23

Rounding Decimals

Quiz 1

Round decimals to the nearest whole number.

1. 2.44

2. 3.76

3. 8.21

4. 9.13

5. 10.87

Quiz 2

Round decimals to the nearest tenth.

1. 3.74

2. 18.16

3. 21.28

4. 293.87

5. 3828.32

Quiz 3

Round decimals to the nearest hundredth.

3.767

18.385

489.424

387.348

2338.542

Subtract Decimals

Quiz 1

1. 84.83-27.2

2. 987.23-876.87

3. 2548.54-2483.67

4. 473.9-364.54

5. 5784.564-283.84

Quiz 2

1. 4933.45-3475.50

2. 3848.5-38.44

3. 4855.78-4843.49

4. 6858.485-473.345

5. 847.49-384.73

Quiz 3

1. 8473.45-384.87

2. 3854.68-2636.876

3. 4635.245-364.384

4. 4383.98-3736.48

5. 5867.32-3947.2

Multiplying Decimals

Quiz 1

1. 78.32×2
2. 854.21×3
3. 698.1×9
4. 767.3×5
5. 1982.45×5

Quiz 2

1. 493.43×3.4
2. 5493.25×20
3. 3629.67×4.5
4. 1982.32×8.1
5. 1972.78×0.32

Quiz 3

1. 73.37×0.81
2. 144.29×8.2
3. 56.34×7.1
4. 865.45×1.1
5. 342.12×6.3

Dividing Decimals

Quiz 1

1. 56.24÷2
2. 67.62÷3
3. 75.2÷4
4. 18.62÷2
5. 144.24÷4

Quiz 2

1. 87.35÷1.5
2. 90.25÷2.5
3. 100.2÷4.2
4. 200.54÷3.2
5. 1,000÷5.5

Quiz 3

1. 87.54÷3
2. 78.21÷7
3. 872.43÷2.1
4. 5628.65÷0.05
5. 526.45÷0.5

Graphing in the Coordinate Plane

Quiz 1

1. A(2,3)

2. B(3,4)

3. C(6,4)

4. D(5,5)

5. E(7,1)

Quiz 2

1. A(6,2)

2. B(7,5)

3. C(2,9)

4. D(3,8)

5. E(1,5)

Quiz 3

1. A(0,0)

2. B(0,5)

3. C(4,0)

4. D(6,2)

5. E(5,3)

1 x 1 = 1	2 x 1 = 2	3 x 1 = 3	4 x 1 = 4	5 x 1 = 5	6 x 1 = 6
1 x 2 = 2	2 x 2 = 4	3 x 2 = 6	4 x 2 = 8	5 x 2 = 10	6 x 2 = 12
1 x 3 = 3	2 x 3 = 6	3 x 3 = 9	4 x 3 = 12	5 x 3 = 15	6 x 3 = 18
1 x 4 = 4	2 x 4 = 8	3 x 4 = 12	4 x 4 = 16	5 x 4 = 20	6 x 4 = 24
1 x 5 = 5	2 x 5 = 10	3 x 5 = 15	4 x 5 = 20	5 x 5 = 25	6 x 5 = 30
1 x 6 = 6	2 x 6 = 12	3 x 6 = 18	4 x 6 = 24	5 x 6 = 30	6 x 6 = 36
1 x 7 = 7	2 x 7 = 14	3 x 7 = 21	4 x 7 = 28	5 x 7 = 35	6 x 7 = 42
1 x 8 = 8	2 x 8 = 16	3 x 8 = 24	4 x 8 = 32	5 x 8 = 40	6 x 8 = 48
1 x 9 = 9	2 x 9 = 18	3 x 9 = 27	4 x 9 = 36	5 x 9 = 45	6 x 9 = 54
1 x 10 = 10	2 x 10 = 20	3 x 10 = 30	4 x 10 = 40	5 x 10 = 50	6 x 10 = 60
1 x 11 = 11	2 x 11 = 22	3 x 11 = 33	4 x 11 = 44	5 x 11 = 55	6 x 11 = 66
1 x 12 = 12	2 x 12 = 24	3 x 12 = 36	4 x 12 = 48	5 x 12 = 60	6 x 12 = 72

7 x 1 = 7	8 x 1 = 8	9 x 1 = 9	10 x 1 = 10	11 x 1 = 11	12 x 1 = 12
7 x 2 = 14	8 x 2 = 16	9 x 2 = 18	10 x 2 = 20	11 x 2 = 22	12 x 2 = 24
7 x 3 = 21	8 x 3 = 24	9 x 3 = 27	10 x 3 = 30	11 x 3 = 33	12 x 3 = 36
7 x 4 = 28	8 x 4 = 32	9 x 4 = 36	10 x 4 = 40	11 x 4 = 44	12 x 4 = 48
7 x 5 = 35	8 x 5 = 40	9 x 5 = 45	10 x 5 = 50	11 x 5 = 55	12 x 5 = 60
7 x 6 = 42	8 x 6 = 48	9 x 6 = 54	10 x 6 = 60	11 x 6 = 66	12 x 6 = 72
7 x 7 = 49	8 x 7 = 56	9 x 7 = 63	10 x 7 = 70	11 x 7 = 77	12 x 7 = 84
7 x 8 = 56	8 x 8 = 64	9 x 8 = 72	10 x 8 = 80	11 x 8 = 88	12 x 8 = 96
7 x 9 = 63	8 x 9 = 72	9 x 9 = 81	10 x 9 = 90	11 x 9 = 99	12 x 9 = 108
7 x 10 = 70	8 x 10 = 80	9 x 10 = 90	10 x 10 = 100	11 x 10 = 110	12 x 10 = 120
7 x 11 = 77	8 x 11 = 88	9 x 11 = 99	10 x 11 = 110	11 x 11 = 121	12 x 11 = 132
7 x 12 = 84	8 x 12 = 96	9 x 12 = 108	10 x 12 = 120	11 x 12 = 132	12 x 12 = 144

Scan Here for Additinal Resources

Scan Here for Answer Keys

www.ingramcontent.com/pod-product-compliance
Lightning Source LLC
LaVergne TN
LVHW052027200926
848481LV00005B/1607